Evolution: The Golden Calf

by
Dean R. Zimmerman

HAWKES PUBLISHING, INC.

3775 S. 5th W. (Box 15711)
Salt Lake City, Utah 84115

Copyright 1976
by Hawkes Publishing Inc.

ISBN 0-89036-059-6

First Printing, February, 1976

Printed in the U.S.A.

TABLE OF CONTENTS

PREFACE

Evolution: a Golden Calf, was prepared with the idea hat academic freedom is not fostered by default— that truth is arrived at through a process of dialogue not monologue.

Too often in an area where two conflicting points of view exist there are those who would say, "We'll stand back from the academic exchange, hoping 'our side' emerges victorious." But all too often because of complacency, perhaps because of ignorance, the "other side" goes unchallenged, and truth loses the battle.

Though written firmly, what is here written is tentative, but so also are the theories of evolution. What must be learned by all is that ALL the answers are not now known. The final results cannot be tabulated. We simply do not have the final word—and with our limited knowledge of the universe and its laws, even if we had it, we probably wouldn't understand it.

Thanks must go to Robert W. Bass and James A. Widenmann, whose friendship and assistance have encouraged this publication; to my students who provided the motivation; and to my wife who was the "sounding board" for my thoughts.

Chapter One
EVOLUTION A FACT?

In 1859 a small green book entitled *On The Origin of Species by Means of Natural Selection, or The Preservation of Favored Races in the Struggle for Life* was published. Its author: Charles Darwin. One modern text said that *"The Origin of Species* was to prove one of the most significant books in scientific literature, revolutionizing concepts about the origin and evolution of life on this planet."[1] And that it did! Not because the work was without flaw, but because at last someone was able to talk convincingly about a theory that had been kicking around since ancient Greece.

Since 1859 a virtual avalanche of books, monographs, papers, theses and articles have spilled from the pens of the layman as well as the learned. Many heralded Darwin's works[2] as the ushering in of the millennial reign of science; others, more cautious, nevertheless considered it a "new wave" of the future in science; while still others thought it depraved at best, and certainly of the devil.

The teaching of evolution in our public schools has raised a few eyebrows wherever it was done, but in Tennessee—the scene of the famous Scopes trial—a law had been passed which forbade the teaching of evolution in that state. It was not until 1968 that the United States Supreme Court ruled the anti-evolution law in Tennessee unconstitutional. The "monkey war" was finally over.[3]

What the Supreme Court, as well as many others failed to realize or acknowledge was that the "war" had ended for the evolutionists nine years earlier, in 1959. Many evolutionists had considered the theories presented by Darwin, Lamarck, Agassiz, Huxley, Spencer and Lyell something more than mere theories, but at the centennial celebration of Darwin's *Origin of the Species* they announced to the world that they had decided that they were to talk of Darwin's work no longer as a theory but as a LAW.

The full impact of this declaration can only be appreciated when it is understood that these learned evolutionists made scientific history. For the first time—ever—a theory of science had been elevated to the status of law without verification of the evidence, without the painstaking collaboration, comparison, checking and re-checking with those with different views and theories that has been the rule since the scientific method came forth from swaddling clothes. Perhaps the day will come when more of the scientific community will recognize the year 1959 as the year that science took a giant step BACKWARDS.

In the past few years, needless to say, the evolutionist presses have been running at top speed, cranking out statements declaring the validity of the "law" of evolution.

As part of a series for "The Thinker's Library," H. G. Wells, in his work entitled *A Short History of the World,* declared, "One thing is certain. No one fact has ever emerged, in a stupendous accumulation of facts, to throw a shadow of doubt upon what is still called the 'Theory' of organic evolution. In spite of that violent lying and yapping of the pious, no rational mind can question the invincible nature of the evolutionary case."[4]

At the conference of 2,500 assembled delegates at the Darwinian Centennial, Sir Julian Huxley affirmed, "We all accept the fact of evolution. . . . The evolution of life is no longer a theory. It is a fact. It is the basis of all our thinking."[5]

The school textbook *Biology For You* says, "All reputable biologists have agreed that the evolution of life on the earth is an established fact."[6]

A university president recently proclaimed, "It takes an overwhelming prejudice to refuse to accept the facts, and anyone who is exposed to the evidence supporting evolution must recognize it as an historical fact."[7]

Geneticist Theodosius Dobzhansky, in his book *Mankind Evolving* tells us that evolution is "verified as fully as an event of the past not witnessed by observers capable of recording and transmitting their testimony can be. ..."[8]

German professor Ernst Haeckel declared, "There is no doubt that man is descended from an extinct mammalian form, which, if we could see it, we should certainly class with the apes. ... It is equally certain that this primitive ape in turn descended from an unknown semi-ape, and the latter from an extinct pouched animal."[9] He then describes the process of evolution in terms so graphic one might think he was reading from his own journal entry of the day under consideration. He said,

> When animated bodies first appeared on our planet, previously without life, there must, in the first place, have been formed, by a process purely mechanical, from purely inorganic carbon combinations, that very complex nitrogenised carbon compound which we call plasson, or 'primitive slime,' and which is the oldest material substance in which vital activities are embodied. In the lowest depths of the sea such homogeneous amorphous protoplasm probably still lives in its simplest character, under the name of bathybius. Each individual living particle of this structureless mass is called a monera. The oldest monera originated in the sea by spontaneous generation, just as crystals form in the matrix. He however, who does not assume a spontaneous generation of monera to explain the first origin of life upon our earth, has no other resource but to believe in a supernatural miracle; and this, in fact, is the questionable standpoint still taken by many so-called "exact naturalists," who thus renounce their own reason.[10]

Statements by reputable scientists that evolution is a law would all but overwhelm us if it were not that in the scientific community evolution is an enigma, a paradox at best.

The father of modern evolutionary theory, Charles Darwin, admitted in 1859 that "long before the reader has arrived at this part of my work, a crowd of difficulties will have occurred to him. Some of them are so serious that to this day I can hardly reflect on them without being in some degree staggered."[11] Many scholars today face similar perplexities when faced with both theories of evolution and the evidence of their investigations.

Science Year of 1966 admitted, "Archaeology, despite its triumphs, remains almost at the beginning of the immense task of reconstructing mankind's history."[12]

Dobzhansky, in his book *The Biological Basis of Human Freedom,* said first that "evolution as a historical fact was proved beyond reasonable doubt not later than in the closing decades of the nineteenth century; . . ." but two pages later says, "There is no doubt that both the historical and the causal aspects of the evolutionary process are far from completely known. . . . The causes which have brought about the development of the human species can be only dimly discerned."[13]

The *Encyclopedia Britannica* explained to its readers, "We are not in the least doubt as to the fact of evolution. . . . The evidence by now is overwhelming." Eleven pages later it admitted that the "overwhelming" evidence is very imperfect and often interrupted by gaps.[14] It then tells that "Of the vital processes which brought about these changes we are as yet ignorant."[15]

Evolutionist and biographer Sir Gavin de Beer in his work, *Charles Darwin* writes, "He [Darwin] predicted that the evidence would one day be forthcoming, and that day has arrived, for the series of fossils just mentioned provides the crucial evidence that man did evolve."[16] And yet the evidence was nowhere to be found in 1964 when evolutionist W. Le Gros Clark published his work, *The Fossil Evidence for Human Evolution.* Clark observed that "the chances of finding the fossil remains of *actual* ancestors, or even representatives of the local geographical group which provided the actual ancestors, are so fantastically remote as not to be worth consideration."[17] He continued later in his book by saying, "The interpretation of the paleontological evidence of hominid evolution which has been offered in the preceeding chapters is a provisional interpretation. Because of the incompleteness of the evidence, it could hardly be otherwise."[18]

In 1965 *Science* magazine reviewed the book *The Basis of Human Evolution* and declared, "The reader . . . may be dumbfounded that so much work has settled so few questions."[19]

L. M. Davies reported that, "It has been established that no fewer than 800 phrases in the subjunctive mood (such as 'Let us assume,' or 'We may well suppose,' etc.) are to be found between the covers of Darwin's *Origin of Species* alone."[20] Some evolutionist scholars still speak that way. To illustrate, a former president of the American Association for the Advancement of Science, whose purpose was to strengthen the position of evolution said in part,

> *Come, now, if you will, on a speculative excursion into prehistory. Assume the era in which species* **sapiens** *emerged from the genus* **Homo** . . . . *Hasten across the millenniums for which present information depends for the most part on conjecture and interpretation to the era of the first inscribed records, for which some fact may be gleaned.*[21]

Dr. T. N. Tahmisian, physiologist for the Atomic Energy Commission, said, "Scientists who go about teaching that evolution is a fact of life are great con men, and the story they are telling may be the greatest hoax ever. In explaining evolution we do not have one iota of fact."[22]

To quote J. W. Klotz, the "acceptance of evolution is still based on a great deal of faith."[23]

In spite of the contradictory conclusions which scientists and scholars reach one cannot read a modern text on the subject of biology, anthropology, geology, sociology, psychology, astronomy, history, or religion without encountering the doctrines of evolution. It is virtually impossible to view a television program dealing with any of these topics without having the subject interpreted in terms of evolution. But considering the contradictory statements of men whose scientific credentials are impeccable, should we not critically evaluate the evidence on which is based the theories of evolution?

Should we not be apprised of those statements which are factual and those which are merely subjective inferences?

Is Evolution a Golden Calf?

Chapter Two

EVOLUTION IN
HISTORICAL PERSPECTIVE

Evolution is a law of science, the evolutionists say. Everybody knows that. But what they don't say is that upon close examination the observer finds to his amazement that what is thought to be "evolution" has suddenly become not one, but two contradictory types of theories.

Specifically, one type of *organic* evolutionist is one who believes that life originated on this planet spontaneously, and by a series of mutations (or other small genetic changes) "simple" life has arrived at its present state of complexity. It involves proceeding from organic simplicity to complexity, from generalized to specialized.

The other type believes as does John A. Widtsoe, Frederick J. Pack and a host of others that evolution means simply a change—either progressive, recessive, or retrogressive.

Biologist Widtsoe said that evolution in its widest meaning "refers to the unceasing changes within our universe. Nothing is static; all things change. Stars explode in space; mountains rise and are worn down; men are not the same today as yesterday. . . . The face of nature has been observed by man from the beginning, and must be accepted by all thinking people."[1]

Pack explained, "Stripped of all complexity and viewed in its simplest form, evolution is scarcely more than the principle of cause and effect. Its operation involves at least two phenomena, that which acts and that which is acted upon."

Scientist Pack further stated,

In a slightly more technical sense the term evolution means "to roll out of" or "to unroll like a scroll," and, therefore, it consists of the sequential stages through which the thing has passed. While evolution is commonly thought of as being progressive, variations may take place in either direction, progressive or retrogressive.[2]

Anthony Standon, in speaking of this broad definition of evolution, which he called "Vague Evolution," said,

Vague evolution is rather difficult to formulate, because it is vague, but it is extremely easy to see. Any book on biology is full of it, and it has been so thoroughly popularized that there is hardly anybody who is not aware of it. It points to the striking similarities, in every detail, between the bodies of men and of the apes; to the slightly more distant resemblances between men and other mammals, to the duck-billed platypus, which Huxley called "a museum of reptilian reminiscense," to the reptiles themselves, to the fish, both bony and cartilaginous, and so on and so on, as can be found in many a fine book. It points, too, to the development of the embryo, "climbing up the family tree," and to the record of the rocks—there were fish before there were reptiles, reptiles before mammals. Whatever this proves—and it would seem to prove that all forms of life are connected in some way—is indisputable.[3]

But in what way? To answer this question, we need a precise theory.

Standon has more to say as to just what he thinks the "precise theory" is, and we shall return to him shortly.

The first definition of evolution—the "all-pervading principle of cause and effect," the "rolling out" theory of "vague evolution," cannot be refuted, it would seem. One thing is certain. Because it is so vague, so intangible, there is nothing with which the scientists can work.

If evolution were "cause and effect," is it only cause and effect? If it were, then why is there the intimidation, browbeating, and harrassment of the scientists who choose not to believe in the evolutionary hypothesis?[4] If evolution were merely the "unfolding" of the history of the world, then why not make history a science and the historians scientists. The fact of the matter is that the evolution which Darwin's disciples preach involved more than this broad definition. A precise theory is necessary.

"Precise Evolution" has been defined in the *Houston Post* this way. "Evolution, in very simple terms, means that life progressed from one-celled organisms to its highest state, the human being, by means of a series of biological changes taking place over millions of years."[5] Another source defined it, "When living things came out of the sea to live on land, fins turned into legs, gills into lungs, scales into fur."[6]

Henry M. Morris explained,

> *Evolution does not simply mean change. This is important, because the evidence cited by most writers in favor of their claim that evolution is a **fact** is simply evidence of change. But true evolution is a certain kind of change.*
>
> *Once again, we shall let evolution's chief present-day spokesman and protagonist, Sir Julian Huxley, settle this particular question:*
>
> *"Evolution is a one-way process, irreversible in time, producing apparent novelties and greater variety, and loading to higher degrees of organization, more differentiated, more complex, but at the same time more integrated." This statement was intended to include both inorganic and organic evolution, and to comprehend the whole of the physical and biological universes. That is, everything in the universe has been developed by this process of evolution, of development, or progress, of higher and higher levels of organization and complexity.[7]*

World Book Encyclopedia explained:

> *The theory of organic evolution involved these three main ideas: (1) Living things change from generation to generation, producing descendants with new characteristics. (2) This process has been going on so long that it has produced all the groups and*

kinds of things now living, as well as others that
lived long ago and have died out, or become extinct.
(3) These different living things are related to each
other.[8]

Evolutionist Kellogg maintains that evolution is an
"outrolling," an "unfolding" of life; but declares that the
progression is only in one direction. He wrote,

> Organic evolution is the outrolling of the plan of
> life. It runs naturally and logically from simple to
> complex, from the general to the special, from the
> lowly to the high, from amoebas—and simpler—to
> man. . . . Evolution means continuity, means
> transmutation, the origin of the new from the old;
> means change, continuous movement, gradatory
> development. It means genetic relationship, blood
> cousinhood, an all-embracing, genealogy of life. . . . It
> means the fundamental unity of all life. It means a
> continuous living stream varying in appearance in its
> different parts, but never really broken or with its
> parts really separated. . . .
>   The theory of descent (which we phrase organic
> evolution may be practically held as a synonym) is
> then, simply the declaration that the various living as
> well as the now extinct species of organisms are
> descended from one another and from common
> ancestors."[9]

Evolutionist Bateson, in speaking of the evidence for
evolution remarked, "It is easy to imagine how Man was
evolved from an Ameba, but we cannot form a plausible
guess as to how *Veronica agrestis* and *Veronica polita* were
evolved, either one from the other, or both from a common
form. We have not even an inkling of the steps by which a
Silver Wyandotte fowl descended from *Gallus bankiva*, and
we can scarcely even believe that it did."[10]
    J. Arthur Thomson explained that "many of the
genealogical trees which Haeckel was so fond of drawing
have fallen to pieces. Who can say anything except in a
general way, regarding the ancestry of Birds or even
Vertebrates? The *Origin of Species* was published in 1859,

but who today has attained to clearness in regard to the origin of any single species?"[11]

In explaining Haeckel's drawings of modern man's genealogical heritage, *Life Magazine* apologetically wrote, "THE ANCESTRY OF MAN, traced by naturalist Ernst Haeckel in 1867, was one of the first attempts to deal with the specifics of evolution. Although his genealogical chart, starting with a blob of protoplasm and continuing to a modern Papuan, is filled with *misconceptions* and *fictitious creatures*, it is *fairly accurate*, considering the dearth of knowledge in his day."[12]

The theory, or "Law" of evolution did not, of course, originate with Charles Darwin. We learn from histories that the ancient Greeks taught of evolution. Anaximander taught that men had evolved from fish. Empedocles theorized that animals evolved from plants. Evolutionist P. Amos Moody explained, "Ideas that by one means or another evolution does occur far antedated Darwin, however. In fact, such ideas are probably as old as human thought."[13] Since the Greeks, others too have interpreted history and life in terms of evolution; but no other theory has ever caught on as has Darwin's.

Probably the first "modern" theory of evolution to gain acceptance was propounded by English naturalist Erasmus Darwin, the grandfather of Charles Darwin, and the French scientist Comte de Buffon, who taught that when a plant or an animal acquired a new characteristic from its environment it could pass this on to its offspring. If there occured a localized stimulation upon, say, the belly of a worm, there would be produced a blister or a scab. Over eons of time this scab, by being subsequently passed on to its posterity, could very conceiveably (with this theory) develop or evolve into an arm. Eyes were formed by the localization of light. Fingers were formed by the constant irritation of the tips of arms. Ears were developed by the constant bombardment of sound to certain positions on one's head.

Early in the nineteenth century Jean de Lamarck had written concerning the Darwin-de Buffon theory of acquired and inherited characteristics. He agreed with the theory, but thought it to be too passive, and that the acquisition of characters needed a force to impel it along to its present state. According to his theory giraffes got long necks because they ran out of easily accessible vegetation and had to stretch their necks to obtain food higher up the tree limbs. Some animals developed teeth, others evolved odors, a few produced offspring that could change colors, and others evolved tough skin. Lamarck taught that all nature must arm itself with whatever it thought needful in order to survive its environment. Thanks to him, modern evolution has behind it a force—Survival of the Fittest.

Sir Gavin De Beer recounts that, "Nobody would have thought of doubting it 'till the close of the nineteenth century . . . . The number of men before the nineteenth century who rejected the inheritance of acquired characters could be counted on the fingers of one hand."[14] But by the end of the nineteenth century the theory of acquired characteristics had been almost completely rejected. German scientist August Weismann tested the theory of acquired characteristics by cutting the tails off 20 successive generations of mice. Hall and Lesser explained that,

> This was the first experimental proof that acquired characteristics, such as artificial taillessness, are not inherited. . . .
> Acquired characteristics. . . , they concluded, . . . are not inherited because environmental factors (which do not affect the genes in the sex cells) cannot influence the next generation.[15]

Geneticist H. J. Muller explained why the theory of acquired characteristics is inadequate, saying,

> Despite the strong influence of the environment in modifying the body as a whole, and even the protoplasm of its cells, the genes within the germ-cells of that body retain their original structure without specific alterations caused by the modification of the body, so that when the modified individual reproduces it transmits to its offspring genes unaffected by its own "acquired characters."[16]

Perhaps the last vestige of this now-disproved theory was in Russia where as late as 1948 under the leadership of

Lysenko the theory of "acquired characteristics" was taught as a fact. Since then Lysenko has been removed from office, and Communism now admits that heredity "is controlled by genes in the reproductive cells and remains unchanged throughout an individual's life."[17]

The longest-enduring theory of evolution came from the work of Charles Darwin. His theory postulated that members of different species compete with each other for survival, and that in the struggle any competitor who possessed an advantageous variation would enable its possessor to gain the upper hand. The fittest therefore would survive. The others would perish. The phrase "Natural Selection" was used by Darwin to identify this process although the phrase "Survival of the fittest" coined by Lamarck is today used interchangeably with it.

Some scholars claim that there are many problems associated with Darwin's theory. Clark and Mould, in their text *Biology for Today*, write that, "the theory does not account for all the known facts of heredity. For example, the theory does not clearly explain why some variations are inherited and others are not. Many variations are so trivial that they could not possibly aid an organism in its struggle for existence." They also believe that "the theory does not explain how the gradual accumulation of trivial variations could result in the appearance of some of the more complex structures found in higher organisms."[18] Another evolutionist, H. Mellersh, very clearly exposed one serious drawback to any theory of evolution for survival. He explained,

> On the Darwinian theory, the questioner may point out, any variation has to be of **immediate** value to its possessor if it is going to give him a better chance of survival than his fellows. Of what "survival value" is the first dim beginnings of an eye, or forelimbs starting to flap about feebly and nakedly in anticipation of a wing? . . . Natural Selection is so **mindless**. It is so purposeless.[19]

Charles Darwin's theory that an advantageous variation (a mutation—though he did not refer to it as such) would allow the possessor to survive by means of a process which he called Natural Selection, gave way to another explanation of the origin and perpetuation of life in the year 1901.

At the turn of the century evolutionist Hugo De Vries was experimenting with Evening Primrose plants. He had observed that primrose plants occasionally develop freakish characteristics which would in part or in total be transmitted to its offspring. In 1901 he formulated his theory that favorable large mutations accounted for evolution.

As with Darwin's theory, De Vries theory (1) does not account for all the known facts of heredity, (2) it does not explain trivial variations, and (3) the variations must be of immediate survival value.

Evolutionist de Beer pointed out that with De Vries experiments,

> Many of them had lethal results and killed the organisms that carried them, . . . Far from conferring improvement in adaptation, the mutations seemed to be pathological, and provided no explanation of how adaptations arose and became perfected. The result. . . was that during the first twenty years of the twentieth century evolutionary studies and theories were in a state of chaos and confusion.[20]

De Vries' experiments with mutant Evening Primrose plants have accounted for a great deal of popular misunderstanding about evolution's supposed experimental basis. What would be understood is that the primrose remains to this day a primrose. It has not evolved into anything else. Subsequent generations vary from the parent plants but variation is not evolution of the "precise" type.

Mutations do account for change—both for survival and death. Hemophilia mutations, for example, occur at the rate of about 1 in 50,000 births. That certainly is not productive. It's not evolution of the "precise" type. Yet when one considers mutations, he must also consider the problems: firstly, statistical improbability of a mutation contributing to the survival of the organism; and secondly, the staggering number of mutations which must have occurred simultaneously in order for it to have contributed to the survival of the organism.

Henry M. Morris reminds us that,

When one considers the great odds against a mutation's being helpful and surviving in the struggle for existence and then realizes that the formation of a new species would require not one mutation but thousands, and finally considers the tremendous number of species of plants and animals in the world, it would seem to demand a most amazing credulity to imagine that here is the method by which evolution takes place. And yet that is precisely what is taught as gospel truth in probably the majority of schools today.[21]

What is Evolution? Is it a Golden Calf molten out of trinkets and rings, fashioned by some for their own entertainment?

Chapter Three

MAN FROM APE?

Some evolutionists teach that man came from ape, although some of them say it another way: that between ape and man there is a transitional animal—a missing link. *Science News Letter* of May 29, 1965, describes our ancestors as:

> hairy, tailless, and a little larger than present-day gibbons. They had mobile facial muscles and no "mental enimence." . . .
>
> They were expert climbers and spent much of their lives in trees. On the ground they could stand with a semi-upright posture. They could walk on all fours and could run on their feet. . . .
>
> The proto-hominoids apparently did not have the power of speech.[1]

And while these evolutionists maintain that man came from ape, and they build their theories of the origin and perpetuation of life upon that premise, few have been as definitive in explaining the step-by-step, graudal development of man from the first form of life as was Ernst Haeckel. In his work *History of Creation*, Haeckel lists 21 (hypothetical) steps from "protoplasm" to man. Joseph Hassell, Associate of King's College, London, listed and then commented on them.

Step 1. Minute portions of structureless protoplasms—the monera of today—"Organisms without Organs." In the course of time, by differentiation an inner kernel was developed, and thus there was produced—

21

Step 2. Single-celled creatures, like the amoeba of the present day. In the process of time these primordial creatures became sponges.

Step 3. These associated amoebe gave birth to ciliated larva, which, by natural selection, produced a new race of beings, viz.:

Step 4. Simple-stomached animals?primitive worms which, after untold ages, gave rise to—

Step 5. Gliding worms, which, not being content we must suppose with their lowly estate, determined to improve their condition, and so gave birth to—

Step 6. Soft worms—the scolecida. These creatures, by some unaccountable means, formed for themselves a true body cavity, and managed somehow or other—the professor does not say how—to possess blood. In the course of ages these soft worms gave rise to—

Step 7. Sack-worms, which originated out of the former creatures by the formation of a dorsal nerve, and by the formation of a spinal rod, which lies between it. After many ages these creatures produced—

Step 8. Skulless animals like the present lancelet. These wise animals managed to produce a progeny in which the sexes were separate. In the course of time these creatures gave birth to quite a different race altogether, and thus were formed—

Step 9. Single nostrilled animals, which were developed out of the former by the anterior end of the dorsal marrow forming itself into a brain, and the chord into a skull. In the course of ages these creatures evolved themselves into—

Step 10. Primaeval fish. In these animals the nostril divided itself; a double nervous system was evolved; jaws were formed; a swim-bladder made its appearance; and two pairs of legs were developed; and so was produced—

Step 11. The mud-fish, somewhat like the present salamander, and this was effected by the adaptation of life on land. The swim-bladder was now made into an air-breathing lung, and thus was produced—

Step 12. Gilled amphibians, such as are met with in the present day. In the course of ages these gave birth to—

Step 13. Tailed amphibians. These creatures accustomed themselves to breathe only by means of gills in the early stages of their life, and in the latter stages through lungs. In the course of ages these gave birth to—

Step 14. The primaeval amniota. These were evolved out of an unknown tailed amphibian, by the loss of gills. Strange to say, the organs of tears were now developed. How wonderful! After many ages these creatures were evolved into animals with hairs and mammary glands, and so—

Step 15. Primary mammals, closely related to the ornithorhynchus of the present day, were produced. By degrees these monotremata produced—

Step 16. Pouched animals. In the course of time one of these marsupial creatures produced—

Step 17. Semi-apes, which, in the lapse of ages, produced the animals of the narrow-nosed monkey tribe, and out of these were evolved—

Step 18. The tailed apes of the New World, which, in the course of ages, produced—

Step 19. The man-like apes (anthropoides) which, in the process of time, lost their tails and a portion of the hairy covering on the back. Poor things! How much inconvenience they must have suffered on this account! When speaking of these creatures the professor says, "There do not exist direct human ancestors among the anthropoides of the present day, but they certainly existed among the unknown extinct human apes of the Miocene period." We beg the reader to mark this assumption,—"they certainly existed"—that is, they existed in the professor's imagination. In the face of this assumption, however, Professor Haeckel continues his steps in the development of man as if it were a thing of certainty, and states that in the process of time these man-like apes produced—

Step 20. Ape-like men. In the course of time out of these were evolved—

Step 21. Man, who has developed out of the former race by the gradual development of the brain and the larynx, so that language and mental power were the result. All these changes were produced by natural selection, resulting in "the survival of the fittest."

Hassell continues by saying,

> Such is the creed of the learned professor, and such must be, he says, the creed of every man who claims to be scientific. "We must," writes the professor, "either accustom ourselves to the idea that all the various species of animals and plants, man also included, originated independently of each other by the supernatural process of a divine creation—or we are compelled to accept the theory of descent in its entirety, and trace the human race, equally with the various animal and plant species, from an entirely simple primaeval parent form. Between these two assumptions there is no third course; either a blind belief in creation, or a scientific theory of evolution."[2]

In spite of what Haeckel so boldly declared, all evolutionists do not agree. Evolutionist Jean Rostand remarked,

> We are still arguing, and doubtless will continue for a long time, about the real connection among all these forms. . . . Did man descend from an ape resembling the anthoropoids we know today? Or from an inferior ape? Or even from a primate which did not as yet deserve the name of ape?[3]

Several years ago Scientific American carried an article which in part said,

> Primatologists may therefore be forgiven their fumblings over great gaps of millions of years from which we do not possess a single complete monkey skeleton, let alone the skeleton of a human forerunner. . . . We have to read the story of primate evolution from a few handfuls of broken bones and teeth. Those fossils, moreover, are from places thousands of miles apart on the Old World land mass. . . .
> In the end we may shake our heads, baffled. . . . It is as though we stood at the heart of a maze and no longer remembered how we had come there.[4]

Even evolutionist Julian Huxley, in his book Evolution as a Process reminded his colleagues,

In the great majority of cases the descriptions of the specimens that have been provided by their discoverers have been so turned as to indicate that the fossils in questions have some special place or significance in the line of direct human descent, as opposed to that of the family of apes. It is. . . unlikely that they could all enjoy this distinction. . . .

In the case of primate evolution the inferences are sometimes very insecurely based because of inadequacies of the evidence.[5]

The *Saturday Evening Post* of December 3, 1966, reported, "Investigators. . . have yet to trace the origins of the human line."[6] *New Scientist* confirmed, "The unmistakable correspondence between man and anthropoids points clearly to a common ancestor. But it has not been found and we may have some difficulty in recognizing it."[7] A recent evolution text admitted, "Unfortunately, the early stages of man's evolutionary progress along his individual line remain a total mystery."[8] And even more recently *Scientific American* declared, "The nature of the line leading to living man. . . remains a matter of pure theory."[9] An article in the New York *Times* a short time ago said, "Even today surprisingly little is known of man's own family tree. . . . There are still enormous gaps."[10]

The weight of these reports is staggering. The impact of what has been said and printed that incorporates the doctrine and dogma of the theories of evolution, in spite of the fact that there is so very little evidence, is staggering.

With the realization of the "enormous gaps" in the fossil record it is no wonder that *Scientific American* recently declared: "Pending additional discoveries it may be wiser not to insist that the transition from ape to man is now being documented from the fossil record."[11]

In spite of the evidence, Carl O. Dunbar, Yale geologist, insists that "although the comparative study of living animals and plants may give very convincing evidence, fossils provide the only historical, documentary evidence that life has evolved from simpler to more and more complex forms.[12] Evolutionist George Gaylord Simpson, Harvard Anthropologist, asserts, "The most direct sort of evidence on the truth of evolution must, after all, be provided by the fossil record."[13]

If the fossil record is the only historical, documentary evidence, "the most direct sort of evidence" of the LAW of evolution, then evolution by the scientific method is indisputably the least documented, least substantiated law in all science.

Chapter Four

MAN vs. THE APES

Perhaps the most difficult doctrine which evolution has yet to reconcile to religion is the position that man, by the supposition that he evolved into a higher organism from a man-like ape (or ape-like man), is no more than a specialized primate. The implication, of course, is that *all* physical, intellectual and social traits in man can be observed in a rudimental state in apes. While there are some who speak of the social behavior of apes; the human-like mannerisms of dogs; the intelligence of chimpanzees, dolphins, or whales; true religion teaches that man is a unique being in all creation.

It seems unbelievable, in the first place, to conceive of an organism having been generated spontaneously, and is all but overwhelming to suppose that a brain has been organized simply by blind chance. And yet evolutionists wish us to believe such a thing.

When they search for evidence to prove their hypothesis they come up empty-handed. Australian Geologist Arthur N. Fields declared, "What is evolution based upon? Upon nothing whatever but faith, upon belief in the reality of the unseen—"[1] and Professor H. H. Newman, Univeristy of Chicago, stated, "Reluctant as he may be to admit it, honesty compels the evolutionist to admit that there is no absolute proof of organic evolution."[2]

Because of man's brain and the size of the cranium, evolutionists expect us blindly to accept the "law" of evolution. Such an expectation is preposterous. Evidence abounds which declares that man IS something special. An examination of man's brain, his unique ability with language, and

27

man's intrinsic morality should be sufficient to demonstrate that man is not simply the latest in a long succession of transitional apes. Man is physical, but that does not mean he is "animal." Regarding the brain capacity of man's skull Joseph Hassell explained, ·

> *Evolution will not account for the brain capacity of man's skull. The **average** internal capacity of the cranium in the different races of men has been found to be as follows:—The Teutonic family, 94 cubic inches; the Esquimaux, 91; the Negroes, 85; the Australian, 82; and the Bushmen, 77 cubic inches. Individuals, however, have been found to possess skulls of much larger measurement. But it may be asked, what proof is there that the ancient races of men had equally well-developed brains? We answer all the evidence that is needed. Some time ago a skull was found in the lake dwellings of Switzerland, supposed to have belonged to a man who inhabited that country in what is called the Stone age, and this skull corresponds in size and character with the Swiss of the present day.*
>
> *Another celebrated relic known as the Engis skull, which according to the testimony of Sir John Lubbock, was contemporary with the mammoth, is yet, according to the opinion of Professor Huxley, "a fair average skull, which might have belonged to a philosopher, or might have contained the thoughtless brain of a savage."*
>
> *So much, then, for man. Now, as to the skulls of apes. The adult male orang-outang is quite as large as a small-sized man; a gorilla is larger; yet the former has but 28 inches of brain capacity; the latter only 20 to 34½ inches.*
>
> *Again, the lowest races of men have five-sixths of that of the highest races; while the highest races of apes have scarcely one-third the capacity of man.*[3]

Robert Kuhn, who was awarded the Ph.D in Anatomy-Brain Research (neurophysiology) from the UCLA Department of Anatomy and Brain Research Institute, told that,

1. Man's brain is similar to animal brain, merely continuing the gradual increase in complexity evidenced by all mammals from rat to chimpanzee.
2. All brain research—anatomy, biochemistry, electrophysiology—staunchly proclaims that the human brain is just barely superior to chimp brain, whereas chimp brain is substantially superior to rat brain.
3. Consequently, if the human mind is entirely the product of the human brain, then the human brain can be no more than just barely superior to chimp brain, whereas chimp brain must be substantially superior to rat brain.
4. But chimps and rats have qualitatively the same compulsive "thought" patterns.
5. And the self-conscious human mind is supreme beyond measure—unequivocally distinct and irrevocably dissociated from the stereotyped behavior of chimp.
6. Obviously, the slim superiority of the physical human brain cannot account for this yawning chasm between the uniquely unrestrained human mind and the instinctively automatic animal brain.
7. Therefore a non-physical addition must unite with and augment the human brain, converting it into the human mind.[4]

Science editor Paul A. Zimmerman has noted these seven differences between man and ape:

1. "Only man walks upright on two feet." The apes occasionally take a few steps on two legs, but then go back to all fours.
2. Man has a nose with a prominent bridge and elongated tip, while both of these are lacking in the ape.
3. Man does not have thumbs on his feet as does the ape.
4. Man shows the greatest amount of weight at birth in relation to his adult weight, but still is by far the most helpless.
5. "Man's head is balanced on top of his spinal column," while the ape's head is hinged at the front.
6. Man alone is teachable. It is possible to train an ape to do certain things, as one would train a dog or horse, but he is not educable.
7. Man has red lips formed by an extension of the mucous membrane from the inside of his mouth. Apes do not have lips of this nature.[5]

Two things should be clear to us now regarding man and the apes. First, that the brain capacity of the human cranium far surpasses that of the ape; and second, man has about one-third more unspecific cortex than apes. But the significance of man's brain goes beyond the measurements of the brains of animals. Man, as the above research has pointed out, has far more "intelligence"—far more creative ability—than does even the "highest" of any other animal.

Even the most "primitive" of mankind is immeasurably greater mentally than the "smartest" of apes, dolphins, or other animal from which man is said to have "ascended." There is a vast unbridgable gulf between man and the animal kingdom. And one of the most perplexing problems for evolutionists is to explain how it is that man in just a few short millennia have accomplished more in terms of technology, socialization, and culture than all other life in the hypothetical eons of evolutionary development preceeding man.

That which sets man apart from all earth life is his ability to transform images into highly complex symbols, then mentally supplant the real for the symbolic. Man has language, which involves the transforming of "reality" into symbols, retention of the symbols, and highly significantly, the manipulation of complex symbols for which there is no "real," physical counterpart.

There is, I suppose, evidence that a monkey may "think" about a banana, or that a dog may "think" about a bone. But there is no evidence whatever to indicate that any animal less than man has ever thought of a symbol for which there is no real counterpart. Love, hate, beauty, empathy, forgiveness, ecstasy, nobility, courage, ambition—our lexicons are bursting with words which we use for which there is no physical object. Even mathematics is based a great deal upon non-real symbolization. No one has ever heard of the apes using non-real symbolization to create for themselves a better environment.

In spite of the scientific and historical facts to the contrary presented later in this chapter, evolutionists persist in accounting for language in terms of their theories rather than accepting the scriptures which state that man (Adam) was created by God, with whose help he gave names to all animals (language skills).

Listed below are ten theories which have been used variously to explain the presence of language with man.[6]

Social Pressure Theory

Adam Smith taught that social pressure forces human beings to "utter certain sounds when they meant to designate certain objects."

Onomatopoetic or Echoic Theory (Bow-Wow)

Objects were given names which resembled the sounds which those objects made.

Interjectional Theory (Pooh-Pooh)

Etienne Bonnet Condillac theorized that "under emotional strain or intense feeling we instinctively give utterance to ejaculations or exclamations."

"Phonetic Type" Theory (Ding-Dong)

Muller hypothesized that the theory of roots indicates that language is made up of basic modes of articulate utterance. The sounds "produced fall into certain basic phonetic types which are the roots from which all subsequent language is developed."

Yo-Ho-Ho Theory

This theory postulates that any strong muscular effort results in an attempt at relief by the forcible emission of breath, which sets the vocal mechanism to vibrating.

Gesture Theory

Though Wundt does not insist that articulate language develops out of gesture he does say that "every sensation has its peculiar expression, the result of definite neutral connections between the "receptors" [that would be eyes, nose, ears, the sense of touch] and the "effectors [the muscles]".

Vocal Play Theory

Jespersen taught that "primitive speech. . . resembles the speech of the baby himself. . . . Language originated as play, and the organs of speech were first trained in this singing sport of idle hours."

Oral Gesture Theory

The total theory says that man first gestured with the hands and other parts of the body. And, according to Paget, while acting thus, he began to mimic these gestures by moving various parts of the mouth and area of vocal production. By exhaling air through the oral and nasal cavities the particular position of the tongue, lips, jaws, and palates produced a distinct whispered speech sound.

Social Control Theory

Grace de Laguna speculates that speech began out of the serious business of living rather than the gay existence and hilarity of Jespersen's (Vocal Play Theory). "Speech," she writes, "is the means by which the diverse activities of men are coordinated and correlated with each other for the attainment of common and reciprocal ends."

"Contact" Theory

"The central principle in the theory," states G. Revesz, "is the necessity for 'contact' with one's fellows." He theorizes that the "growth of language was an evolutionary process, developing as the race developed." Steps in this imagined evolutionary process begin with the "cry," develop to the "call," and conclude with the "word."

Divine Origin Theory

There is in actuality an eleventh explanation of the development of language; and that is the *Divine Origin Theory* which teaches that language was taught to Adam—the first man, who was created with the ability to articulate both concrete as well as abstract symbols. Harvard professor of philosophy Susanne Langer came to about the same conclusion, though she accepts evolution, when she said,

> The obvious approach to the problem of the origins of speech, treating speech as a higher form of some animal communication, has always proved to be sterile. Speech is so complex a phenomenon that it probably arose from many convergent traits in a very specialized primate.[7]

For over one hundred years the disciples of Darwin have been pleading that the people of Terra del Fuego were the lowest in the scale, so far as discovered, and their language correspondingly crude. But further investigation shows that they have 32,430 words; over twice as many as Shakespeare used.

The language of some of the tribes of the Congo is described by a missionary as more complex than Greek.

The history of languages shows the same want of evidence for an evolutionary origin. The oldest forms are most complex.

Dr. Otto Jespersen, University of Copenhagen, has written:

> We find that the ancient languages of our family, Sanskrit, Zend, etc., abound in very long words; the further back we go, the greater the number of sesquipedalia. We have seen how the current theory, according to which every language started with monosyllable roots, fails at every point to account for actual fact and breaks down before the established truth of lingustic history.[8]

Professor in language at the University of Paris, D. J. Venryes, observed that,

> Some languages have been proved to be older than others, and certain of our modern tongues are known to us in forms more than two thousand years old. But the oldest known languages, the "parent languages," as they are sometimes called, have nothing of the primitive about them. Differ though they may from our modern tongues, they only furnish us with an indication of the changes which language has undergone, they do not tell us how language originated.[9]

What, then, is the evidence of evolution in languages? There is here, like in other fields of study, no evidence at all—only the idle theories of the wishful thinkers. And while wishful thinking is not necessarily to be exluded from all pursuit of truth, it should not be considered the evidence itself. No true science would base its premises upon it.

Max Mueller has written, "There is one barrier which no one has yet ventured to touch,—the barrier of language. Language is our Rubicon and no brute will dare to cross it. . . . No process of Natural Selection will ever distill significant words out of the notes of birds and animals."[10] And in spite of the tremendous lack of evidence that linguistic evolution actually occurred, and upon no other basis than imagination Charles Darwin taught,

> That primeval man, or rather some early progenitor of man, **probably** used his voice largely, as does one of the Gibbon apes at the present day, in producing true musical cadences—that is, in singing; we may conclude from a **widely-spread analogy** that this power **would have been** especially exerted during the courtship of the sexes, serving to express various emotions, as love, jealousy, triumph, and serving as a challenge to their rivals. The imitation by articulated sounds of muscial cries **might have** given rise to words expressive of various complete emotions.[11]

The Old Testament Prophet Isaiah began a great exhortation, and we shall end this chapter with his words, "Come now, and let us reason together. . . ."[15]

Chapter Five

MORPHOLOGY: EVOLUTION BY ANALOGY

Morphology is the study of forms. And in the case of evolution it means the study of the structures of the forms of living organisms (regardless of their function) from which conclusions can be drawn as to their location in the "tree of life." Evolutionists who use morphology as an argument for the evolutionary development of all living things might, for instance, see the commonality in the structure between the flipper of a whale, the wing of a duck, the fore-foot of a horse, and the arm of man; and then draw conclusions regarding their relative evolutionary order.

The periodical *New Scientist,* told, "The unmistakable correspondence between man and anthropoids points clearly to a common ancestor."[1] The argument of "correspondence" is that of morphology. Evolutionist William Berryman Scott, as professor of Geology and Paleontology at Princeton University, rejected Darwin's explanation of the evolutionary development in nature and taught, "To one who approaches the problem from the study of fossils, the doctrine of natural selection does not appear to offer an adequate explanation of the observed facts."[2] But what did he base his belief on? He explained, "We are shut up to drawing of inferences from what may be learned by comparison."[3] Comparison, analogy, morphology—they are all the same. They draw inferences, or educated guesses about the ancestry of living things from their observation of the similarities they perceive in things dead.

Evolutionist Theodosius Dobzhansky, in his book

Mankind Evolving, tells how he sees man's similarity to the apes:

> . . . The human body is constructed on the same general plan as the bodies of other animals, in an order of increasing similarity with vertebrates, mammals, primates and apes. Every bone in the human skeleton is represented by a corresponding bone in the skeletons of apes and monkeys.[4]

Though interesting as it may be, simply because an inference can be drawn from an observation (for example, the correspondence of spinal structures in man and ape) does not show common ancestry. This fact was pointed out recently in the scientific journal New Scientist. After discussing the "unmistakable correspondence between man and anthropoids" which they told "points clearly to a common ancestor," the periodical then explained that the "common ancestor" has not been found, and that scientists "may have some difficulty in recognizing it."[5]

Though impressed with the striking resemblance between the skeletal structures of man and the apes, Dobzhansky admits however that similarities need not mean common ancestry. He said, .

> In the mid-eighteenth century, Linnaeus included man in his classification of animals and considered man and the anthropoid apes species of related genera—Homo and Simia. This may seem surprising since Linnaeus unheld the doctrine of the separate creation of species [unlike the current theories of evolution], but there was no inconsistency here, since Linnaeus believed is the Great Chain of Being, an idea widely accepted during the eighteenth and a good part of the nineteenth centuries. . . . Nature was viewed as a single, linear series of increasingly perfect forms, from inanimate objects, through simple and complex organisms, to man, and to the spiritual world. Placing men and apes in the same zoological genus did not mean that one evolved from the other or that both descended from common ancestors; it meant only that they were about equal in the level of complexity and perfection of their bodies.[6]

Also countering the theory of morphology is Julian Huxley, perhaps the strongest proponent of the evolutionary movement. He taught that "no amount of purely morphological evidence can suffice to prove that things came into existence in one way rather than another."[7] A Professor of Anthropology in the Museum of Natural Science (Paris) wrote, "Without leaving domain of facts, and only judging from what we know, we can say, that morphology itself justifies the conclusion that one species has never produced another by derivation."[8]

English lawyer Henry R. Kindersley discussed the futility of the evolutionary argument of "comparative anatomy" some years ago. He declared,

> The case of the hare and rabbit affords a simple illustration of the futility of expecting "comparative anatomy" to furnish the missing evidence for Evolution; and at the same time it supplies a convincing example of the immutability of "species." Here are two types of rodents exhibiting such remarkable similarities of structure and "posture"—(see Sir Arthur Keith's address, British Association, 1931, on "posture") that if the case for Evolution rested on structural and postural resemblances then evolutionists would triumphantly declare that "all thinking men are agreed" that Evolution has now passed the stage of theory and entered the happy state of certainty. They would claim this to be a clear-cut case of ascent in the scale of life—or was it, perchance, a case of "degeneration!" Which first saw the light of day, the rabbit or the hare? And which of them claims priority of place in the scale of life?
>
> Now let us exchange the hazy area of plausible appearances for the region of realities. Let us follow the advice of Sir Arthur Keith and Professor Julian Huxley and turn to the "species" of the present as the only guide to the "species" of the past. **Examined as living species,** we find that the hare and the rabbit absolutely refuse to interbreed. Moreover, one of them produces its young blind and naked and the other open-eyed and covered with fur. Under Professor Poulton's definition of "species" the fact of

sterility proclaims these two types of rodents (in spite of cogent appearances to the contrary as judged by comparative anatomy, and also that they are both said to chew the cud!) to be unrelated, separate "species"—each in itself an "interbreeding community" sterile with all others. And this case is just one of the million similar prohibitive obstacles in the shape of "living species" which have faced evolutionists since Darwin launched upon the world his agitating theory of Evolution by "natural selection." For Evolution to succeed, this massed wall of living obstacles must be breached or surmounted, one or the other.[9]

Alexander Stewart expressed his views of the weakness of morphology as a plausible defense of evolution, saying,

To argue. . . that because there is physical similarity there must also be identity of being, is to proceed on the basis of manifest fallacy. We might as well conclude that because the bodies of two men are the same in kind their moral character must also be identical. Have we not what is known in chemistry as isomorphous bodies—bodies which are alike in form and similar in chemical constitution, yet different in their properties? The salts formed by these substances, with the same acid and similar proportions of the water of crystallization, are identical in their form, and, when of the same colour, cannot be distinguished with the eye; magnesia and zinc sulphate may be thus confounded. . . In these isomorphous substances the identity of shape is so complete that they all possess the same crystalline form (octahedron, eight sides). No scientist, however, will presume to say that they are identical in kind or in qualities; or that the one has been evolved from the other. Why then should we be expected to believe that because physical resemblances exist more or less between man and the higher apes, he and they should therefore be one save only in the degree of development.[10]

Chemist Anthony Standen sees analogy apart from

science. He sees the biologist who defends their
evolutionary position with morphology as seeking
"something very much like rhyme." Of biology, which he
said has been taken over almost totally by evolution,

> If one of the reasons for bothering about science is
> its logical thinking, then biology is hardly on the
> map. If you take a course in biology, or read any of
> the textbooks, you will find extremely little that can
> be called scientific in any scientific sense. For there
> is practically nothing there but descriptive facts, and
> facts alone do not make a science. . . .
> The truth is that biologists don't think, at least not
> in the narrow sense of making formal conclusions,
> definitely arrived at from definite premises. Their
> mental processes go by analogy. Analogy is a
> wonderful, useful and most important form of
> thinking, and biology is saturated with it. Nothing is
> worse than a horrible mass of undigested facts, and
> facts are indigestible unless there is some rhyme, or
> reason to them. The physicist, with his facts, seeks
> reason; the biologist seeks something very much like
> rhyme, and rhyme is a kind of analogy. A man's arm,
> the front leg of a horse, the wing of a bat, and the
> flipper of a whale—look at them carefully with some
> imagination, and very clear and beautiful analogies
> appear, for they all have the same bones, in the same
> order, modified in various ways. (The biologists
> would speak of "homology" here, and keep the word
> "analogy" for other kinds of resemblance, but he is
> using words in a special, technical sense.) A human
> fetus shows clear analogies to a fish, and by a more
> vigorous exercise of the imagination a biologist can
> see part of the human ear in the jawbone of a fish.
> This analogizing, this fine sweeping ability to see
> likenesses in the midst of differences is the great
> glory of biology, but biologists don't know it, and
> they praise themselves for the wrong reasons. They
> have always been so fascinated and overawed by the
> superior prestige of exact physical science that they
> feel they have to imitate it, and they solemnly

announce that what they are doing is "framing hypotheses" and "testing" them, in the manner of the physicist. . . .

Biology is one vast mass of analogies, very different indeed from the cold logical thinking of the physicist. In the higher reaches, such as genetics, biochemistry, neurophysiology and other "ologies," biologists do some making of hypotheses and testing them against experiment, although even there they are apt to talk of "understanding in terms of. . .", or of "stressing this, or that, aspect. . . ." In its central content, biology is not accurate thinking, but accurate observation and imaginative thinking, with great sweeping generalizations. "The Unity of Life" is a catch phrase they are addicted to, although this can hardly be regarded as confirmed by experiment, because it is almost impossible to say what it means, if indeed it has any meaning at all.[11]

When some use morphology as an evidence for evolution it is no wonder that F. O. Bower, Professor of Botany, Glasgow University, and at the time President of the British Association, confessed, "At the present moment we seem to have reached a phase of "negation" with respect to the attempts of botanists to trace out lines of evolutionary descent."[12] Evolutionist H. F. Osborn admitted, "Were Darwin alive today he would be the first to modify the speculations and conclusions of 1859."[13] And Dr. D. H. Scott, Professor of Botany, University College, London, said, "For the moment. . . the Darwinian period is past; we can no longer enjoy the comfortable assurance which once satisfied so many of us that the main problem had been solved—all is again in the melting-pot. By now in fact a new generation has grown up that knows not Darwin."[14]

Chapter Six

EMBRYOLOGY: THE STUDY OF THE LIFE CYCLE

Embryology is the study of the segment of the life cycle between conception and maturity wherein the organism is considered to be an embryo. In the human, for example, the embryonic stage is the first three months of life following conception, after which the human organism is called the fetus.

Evolutionists who maintain that embryology demonstrates the correctness of the evolutionary hypothesis teach that an examination of the embryos of various animals, including man, gives clear and unmistakable evidence that the correspondence between all these embryos succeeds in establishing evolution as a law, and as fully verified. Charles Darwin taught:

> With respect to development, we can clearly understand, on principle of variations supervening at a rather late embryonic period, and being inherited at a corresponding period, how it is that the embryos of wonderfully different forms should still retain, more or less perfectly, the structure of their common progenitor. No other explanation has ever been given of the marvelous fact that the embryos of a man, dog, seal, bat, reptile, etc., can at first hardly be distinguished from each other. In order to understand the existence of rudimentary organs, we have only to suppose that a former progenitor possessed the parts in question in a perfect state, and under changed habits of life they became greatly reduced, either

from simple disuse, or through the natural selection of those individuals which were least encumbered with a superfluous part, aided by the other means previously indicated.[1]

Theodosius Dobzhansky echoed,

The celebrated gill arches, which are formed in human embryos and in those of other land-dwelling vertebrates, are also present in embryos of fishes, but in the latter they eventually become the supports of functioning gills. Can one avoid the inference that our ancestors had gills that were used as such? The arteries issuing from the heart of a human or any other mammalian embryo had at first an unmistakable resemblance to the arteries of fishes, then to the arteries of amphibians (frogs, toads, and salamanders), only to be completely rebuilt as they later develop. The human embryo at a certain stage has a tail formed like those of mammalian embryos that have tails as adults. Does this not suggest that our ancestors had tails?[2]

Darwin surmised that although the early embryonic stages of various organisms reveals little that is dissimilar, it is inthe later embryonic stage that changes begin to appear. These changes, he postulates, occurred because the parent organism did not use "the parts in question," or "through natural selection of those individuals which were least encumbered with a superflous part."

Darwin's second reason for choosing embryology to prove his theory was that he felt that natural selection had somehow endowed the offspring with an unexplainable ability to survive its constantly changing yet hostile environment.

The argument for embryology by evolutionists is substantially the same as for morphology; that where there are resemblances between two organisms they must—they maintain—be linked by heredity. Fortunately, however, few scientists—even evolutionists—seriously consider embryology as an argument for evolution. For example, evolutionist Agassiz correctly observed that,

. . . The embryonic conditions of the higher vertebrates recall adult forms of lower vertebrates now living, their own contemporaries, just as much and in the same way as they recall the fossil forms. Shall we infer that because a chicken or a dog, in our own day, in a certain phase of its development resembles in certain aspects a full-grown skate, that therefore chickens and dogs now-a-days grow out of fishes? We know that it is not so, and yet the evidence is exactly the same as that which the evolutionists use so plausibly to support their theory. The truth is, that while a partial presentation of the facts seems to sustain his theory, when taken in their true connexion and fairly stated they destroy it by proving too much. They show that the relations between fossil animals supposed to prove descent, exist also between living animals where they have nothing to do with descent.[3]

Joseph Hassell, Associate of King's College, London, said, "For though the life-germ of each class is the same at first, it does not continue the same throughout its development." He continues by saying,

All the eggs of the vertebrates may begin their development in one way and run on in the same way for a while; but the invertebrata begins in another, and in virtue of their own special potentiality they divide, and sub-divide, and weave in one case a protozoan, in another an insect, in another a mollusk, in another a fish, in another a bird, and in another a mammal, as the case may be; and this they always do, and, as far as evidence goes, always have done. Professor Haeckel, who bases his conclusion on man's descent from the amoeba, on the similarity of the egg-cell of all animals, by a diagrammic represen- tation of the egg cleavage of seven distinct classes really shows that the differentiation is different in each. **Thus, while the parent cell of man, frog, and the amphioxum, presents no appreciable difference, the first cleavage state is not at all the same.** In man the cleavage is dual, while in the frog and amphioxus it is quadruple; and, indeed, the whole of the five

separate developments of the cells are dissimilar. . . .
To adopt the language of Dr. Cook, of Boston, we may
say: "Just as the weaver, when he throws his first
shuttle, has the plan of the whole fabric in his mind,
because he has arranged beforehand the pattern, and
has provided for it in the disposition of his warp, so
there is a well-arranged plan settled before to which
each bioplast works; and, in virtue of this pre-
arranged plan, all creatures produce progeny after its
kind. To each seed is given its own body.[4]

Of particular help to zoologists has been the study of
"topographic anatomy" which no longer considers the structure of the body from the stand-point of isolated organs, but
from that of body regions; i.e. the head, trunk, limbs, etc.
According to one zoologist and anatomist, "The growth of
knowledge of the body layers, affords, in fact the most
remarkable feature in the progress of zoology during the
second century of that science's existence. It provided rich
material for new connections of ideas, to which Darwin and
his contemporaries had been strangers." Says Dr. Albert
Fleishmann,

The new view-points stimulated, on all sides,
assiduous research in the wide field of animal
anatomy. The resulting well-grounded knowledge
soon led to a complete change in ideas, which swept
aside the old wide-spread notion of Darwin's day
that the human body supplied the pattern for all
animals, or, as it used to be said, that the organs of
all members of the animal kingdom correspond to
those of a dissected man (L. Oken); a preconceived
notion which, by encouraging talk of "the ascending
scale" of animal species, has led to great confusion.
In place of this notion, the clear conviction arose that
the invertebrate phyla are, throughout their history,
fundamentally different from the Vertebrata
(including man), just as Cuvier had, with admirable
insight, pointed out between the years 1795 and 1832.
Now, . . . we actually recognize more than a dozen
such groups of fundamentally different types of body
structure, namely: Vertebrata, Arthropoda,

Crustacea, Annendes, Rotatoria, Mollusca, Brachiopoda, Echinodermata, Tunicata, Platodes, Bryozoa, Coelenterata, Protozoa.

Had Darwin lived to witness this advance, he would have abandoned his illusion of a single great genealogical tree for all species of animals. The layman, however, could not formerly, and still cannot today, understand why the genealogical tree and the phyla conceptions are so irreconcilably opposed to each other, because he lacks the comprehensive knowledge of the development phases of all the phyla, which would make this opposition clear to him.

Looking back at the importance of modern day anatomy, Fleishmann reflected,

As compared with the obsolete methods of procedure of 60 to 100 years ago, the modern one has the advantage that it takes into consideration not only fully developed body, bqt also all the stages of its growth, from egg to adult. This comprehensive review shows us that the foundations of the ultimate structure are laid down in the earliest stages of existence, and development proceeds, as if of logical necessity, to the pre-ordained magnitude and final conditions. The same identical sequence of earlier and later life stages repeats itself, in the case of each member of the species, just as if the process of bodily development clung to a rigid track, along which the germinal layer complex was compelled to travel during life, through a definite number of fixed intermediate stages to the appointed end. The course of life of every individual within the phylum traverses a special, native and unchangeable sequence of phases, which finally produces the fully developed body will all its parts. **The wonderful regularity shown by the course of this development forbids the idea that the mode of growth within the phylum ever left one track in order to follow another.** It is clear that, in supposing that existing species had sprung from other species, Darwin was only taking *adult structures into consideration. In any case, Darwin's followers must now suppose that the developments of the germinal layers of earlier species underwent very frequent changes! But modern knowledge of the constancy of development shown by species lends no countenance to this.*[5]

Zoologist Douglas Dewar, speaking of "Genetics and Evolution," declared that "The idea that mutations. . . not only took place, but were caused by unidentified natural forces, is, I submit, fantastic. Some nineteen hundred years ago St. Paul said, 'All flesh is not the same flesh: but there is one kind of flesh of men, another flesh of beasts, another of fishes, and another of birds.' Today I think we can go farther and say that all cytoplasm is not the same cytoplasm: but the cytoplasm of each class of animals differs from that of all the other classes."[6]

It is unfortunate for the theory of evolution, that every time they make a claim that a certain scientific study proved evolution, upon investigation just the opposite is the case. Evolution remains unproved.

Regarding the shaky foundation embryology builds for the theories of evolution, Sir Arthur Keith, former president of the Royal Anthropological Institute, said,

It was expected that the embryo would recapitulate the features of its ancestors from the lowest to the highest forms in the animal kingdom. Now that the appearance of the embryo at all stages are known, the general feeling is one of disappointment; the human embryo at no stage is anthropoid in appearance. The embryo of the mammal never resembles the worm, the fish or the reptile. Embryology provides no support whatsoever for the evolutionary hypothesis.

Embryology is not the answer for evolution. The evolutionists must look elsewhere for evidence of their theory.

Chapter Seven

PALEONTOLOGY: THE "HUNT AND PECK" METHOD

Evolutionists sternly maintain that man evolved from ape; not a modern ape, they say, but the common ancestor to the ape. Elwyn L. Simmons matter-of-factly stated:

> A major feature of biological evolution during the past 70 million years has been the rapid rise to a position of dominance among the earth's land-dwelling vertebrates of the placental mammals (mammals other than marsupials such as the kangaroo and primitive egg-laying species such as the platypus). A major feature, in turn, of the evolution of the placental mammals has been the emergence of the primates: the mammalian order that includes man, the apes and monkeys. And a major event in the evolution of the primates was the appearance 12 million to 14 million years ago of animals, distinct from their ape contemporaries, that apparently gave rise to man.[1]

The fossil evidence for evolution has become probably the greatest single weapon which the evolutionists have been able to grasp in order to maintain their academic stance. Geologist Carl O. Dunbar maintained that "although the comparative study of living animals and plants may give very convincing circumstantial evidence, fossils provide the only historical, documentary evidence that life has evolved from simpler to more and more complex forms."[2] Anthropologist George Gaylord Simpson maintained that,

47

"The most direct sort of evidence on the truth of evolution must, after all, be provided by the fossil record."[3] And the periodical *Natural History* not too long ago observed, "In the century since Darwin's controversial theory first appeared, paleontologists have established a solid foundation for evolution."[4]

With statements like these, it would seem that the fossil evidence could be arrayed before us in such splendor and in such completeness that "no rational mind can question the invincible nature of the evolutionary case."[5] But as one begins to investigate the claims and analyze the evidence, it becomes increasingly apparent that what the evolutionists are doing when they speak of the paleontological evidence is almost entirely wishful thinking.

We begin to read statements like this one by evolutionist Dobzhansky:

> The evolution of the animal kingdom, or the vertebrate phylum or the class or mammals, is usually represented as a branching tree. The base of the tree is the, often hypothetical, common ancestor, the branches the diverging and ramifying descendants, and the twigs the species or groups of species. The tree thus symbolized the **cladogenesis**. . . , i.e., the adaptive radiation, the tendency of the evolutionary stream to become subdivided into numerous branches, only to have most of them become extinct because of failures to keep adjusted to changing environment.
>
> It is usually assumed, by a sort of long established tradition, that the evolution of the hominids must also be portrayed as a branching tree. This leads, however, to a predicament. The several australopithicenes, and the Java, Peking, Ternifine, Heidelberg, Solo, Rhodesia, Neanderthal men, and other forms duly provided with generic and specific names in Latin, are placed on the branches or twigs of the human evolutionary pedigree. But almost every one of these fossils exhibits some peculiarities and specializations absent in modern man and in other fossils. Now, because of exaggerated reverence for the principle of irreversibility of evolution, it is difficult to regard any form a direct ancestor of any

other. Despite its many branches, the trunk of the pedigree would remain hypothetical. This interpretation would make man have many fossil collateral relatives but no progenitors. The derivation of **Homo sapiens,** *then, becomes a puzzle. He must have originated in some country in which no fossils have been discovered—an easily defensible hypothesis since human fossils have been found in only a few places.*[6]

If, as evolutionists surmise, life has increased in complexity in gradual successive steps, then one would certainly expect to find the same sort of gradation in the fossil record. The truth is that no such gradation of fossils has been found. Fossils are not found ranging from simple to complex as should be expected under the theories of evolution; but are, according to Conn, "well differentiated." The earliest fossils, contrary to the theories of evolution, are not generalized, but are all specialized. He explained, "In the earliest records geology discloses, we find not a few generalized types but well differentiated forms, nearly all the sub-kingdoms as they now exist, five-sixths of our orders, nearly an equal proportion of sub-orders, a great many families and some of our present species. All this is a surprise and an unexplained problem."[7] What this means is that there is remarkably little evidence in the fossils that there has been any evolution at all.

Sir Roderick Murchison is reported to have said, "I know as much of nature in her geologic ages as any living man, and I fearlessly say that our geologic record does not afford one syllable of evidence in support of Darwin's theory.[8]

Joachim Barrande, telling of the absence of paleontological evidence that there have been transitional forms between the species that now exist or existed in the past, declared:

> *One cannot conceive why in all rocks whatever and in all countries upon the two continents, all relics of the intervening types should have vanished. . . . The discordances are so numerous and pronounced, that the composition of the real fauna seems to have been calculated by design for contradicting everything which the theories [in evolution] teach us respecting the first appearance and primitive evolution of the forms of life upon the earth.*[9]

G. A. Kerkut, speaking specifically of the supposed evolution of horses, explains a general difficulty of paleontology. He said,

> In some ways it looks as if the pattern of horse evolution might be even as chaotic as that proposed by Osborn for the evolution of the Proboscidea, where, in almost no instance is any known form considered to be a descendant from any other known form; every subordinate grouping is assumed to have sprung, quite separately and usually without any known intermediate stage from hypothetical common ancestors in the early Eocene or late Cretacious.[10]

Joseph Hassell said that, "Evolution by natural selection is not borne out by the testimony of geology, or, in other words, by what the rocks declare as to the succession of life on the earth."[11] Evolutionist William Howells admits, "In fact, we must realize that the finding of early man is still only in its early stages."[12]

Certainly the greatest difficulty with the evolutionary hypothesis is that extremely little has been found to substantiate it. N. J. Berrill in his book, The Origin of Vertebrates, discloses that "Proof may be forever unobtainable, and it may not matter, for here is such stuff as dreams are made on."[13] Evolutionists Noel Korn and Fred Thompson write, in their recent work Human Evolution, that "it is now evident that the fossil record itself is simply not enough; . . ."[14] And when they say that the evidence is "not enough," a detailed examination of the fossil record demonstrates the truth of their conclusion.

The following is a list of "men" who have from time to time been envisioned in man's genealogical tree, together with the actual fossil evidence for its separate taxonomic identity.

Name:	Fossil Evidence:
Aegyplopithecus	third molar
Aeolopithecus	third molar
Africanthropus	small pieces of at least three human skulls
Amphipithecus	a little jaw fragment
Apidium	small fragmented "pieces;" three pre-molars

Atlanthropus mauritanicus . see *Ternifine Man*

Australopithecus .... some teeth; skull

Australopithecus africanus . skull; jaw

A. prometheus .... back part of adult skull; broken jaw of young ape

Brahmapithecus .... left half of lower jaw; two molars; roots of a molar and premolar

Cro-Magnon Man.......... many entire skeletons

Cyphanthropus............ see *Rhodesian Man*

Dryopithecus.............. an abundance of jawbones, teeth; a *humerus* (upper arm bone)

Eoanthropus Dawsoni ...... see *Piltdown Man*

Florisbad Man .... frontal parts of a skull

Fontechevade fragments .... a single occipital (back and base of skull)

Gally Hill Man............ an entire (but modern) skeleton

Giantopithecus............ 53 teeth purchased at a Hong Kong drugstore; three lower jaws

Griphopithecus............ one molar tooth

Heidelberg Man .... a lower jaw with most of its teeth

Homo erectus .... a jaw; an occiput

Homo habilis.............. small pair of *tibia* and *fibula* (lower leg bones); lower jaw; two partial *parietal* (top and sides of the skull); bones of a hand

Homo hiedelbergensis ...... see *Heidelberg Man*

Homo Javensis .... see *Java Man*

Homo kanamensis.......... lower jaw

Homo leakeyi .... a skull top

Homo neanderthalensis ..... see *Neanderthal Man*

Homo rhodesiensis .... see *Rhodesian Man*

Homo soloensis .... see *Solo Man*

Homo Sapiens
 palaeohungaricus......... the occipital bone (back of the head)

Hylopithecus............. one molar tooth

Java Man .... left thigh bone; three teeth; a jaw bone

Javanthropus............. *see Java Man*

Kangera Man............. four broken skulls

Lantian Man .... lower jaw; a tooth; skull cap

Limnopithecus .... skull; part of left jaw with three cheek teeth; part of right jaw with two cheek teeth

Meganthropus............. lower jaw fragment, 12 pre-molars, one molar

Moeropithecus .... two molar teeth

Neanderthal Man.......... various fragmented skulls; lower jaw without teeth

Neopithecus.............. a third molar tooth

Niah Cave Boy .... the "better part of a skull"

Notharctus tenebrosus..... various skeletal fragments

Oligopithecus .... some teeth

Oreopithecus bambolii..... nearly complete but flattened skeleton; bits and pieces

Palaeoanthropus .... *see Heidelberg Man*

Palaeopithecus .... a palate holding all the teeth of one side except the incisors

Palaeosimia .... single molar tooth

Paranthropus .... skull; jaw and one tooth

Paranthropus crassidens.... skull; jaw; pelvis

Paranthropus robustos ..... broken skull; four loose teeth; a palate w/one molar, 4 yrs. later, lower end of upper arm bone, upper end of *ulna*, a hand bone, 2 toe bones, ankle bone

Parapithecus .... lower jaw

Peking Man .... three teeth; fragments of two jaws; parts of several skull walls

Piltdown man .... part of a skull, jaw; a few teeth

Pithecanthropus erectur .... *see Trinal Ape man*

Plesiadapis.............. skull; limb and foot bones; numerous jaws, fragments; and teeth

Plesianthropus
  trasvaalensis............ two-thirds of a skull
Pliopithecus............... lower jaw; skull with facial portions; most of the skeleton
Pondaugia .... one molar tooth
Proconsul .... a crushed skull; left upper jaw with the teeth; broken lower jaw containing most of the teeth; 2 ankle bones
Propliopithecus .... half a lower jaw; teeth
Remapithecus brevirostris .. right half of an upper jaw; two molar teeth; socket of canine tooth; socket of central incisor; root of a lateral incisor
R. harilusis .... fragment of right half of upper jaw holding two teeth
Rhodesian Man............ jaw fragments with teeth; several arm, hand and pelvic and upper arm bones
Saldanha Man............. skull fragments
Sinanthropus pekinensis.... skull fragments
Sivapithecus himalayensis.. greater part of lower jaw fragments and teeth
Sivapithecus indicus .... one premolar; one molar
Sivapithecus Middlemissi... two molar teeth
Sivapithecus orientalis ..... greater part of lower jaw
Smilodectas gracilis .... several complete skulls; "many other bones"
Solo Man .... Eleven fragmentary skulls; two shin bones
Sheinheim Man .... a brain case
Sterkfontein Ape Man ...... see *Paranthropus robustus*
Sugrivapithecus........... left lower jaw with two molars; one pre-molar; roots of molar, pre-molar, canine, and incisor
Sivanscombe Woman .... fragment of skull
Telanthropus capensis...... three scraps of lower and upper jaw; part of the snout; palate; and a bit of the radius (smaller fore-arm bone)

Ternifine Man.............. three jaws; parietal bone

Trinal Ape Man .... a section of brain pan; two molars; piece of thigh; 3 more or less complete pelvises

Xenopithecus.............. part of left upper jaw; three molars

Zinjanthropus .... fragmented skull, upper jaw w/some teeth

After an examination of this lengthy list of the fossil fragments, a few observations seem appropriate. First, the evidence is, except for jaws and teeth, almost entirely lacking. As a matter of fact ALL the fossil evidence for the 70 million years of primate evolution could be stacked in the corner of a small room. Geologist T. K. Callard observed,

> Dr. Darwin's hypothesis demands a long line of diversified forms, evolved by minute successive slight modifications. From the Trias to the Eocene no mammal of any kind is found in the New World nor in the old world from the Eocene to the Purbeck. To say that these multitudinous diversified successive forms may have existed although not one of them has yet been found, is simply conjecture, and must not rank as science. Evolution is an hypothesis founded too much upon conjecture. Professor Huxley speaks about the demonstrative evidence of evolution. There is no demonstrative evidence of evolution. It is a necessary postulate of the doctrine of evolution, that from the highest animal down to the lowest speck of protoplasmic matter in which life can be manifest there must be a series of gradations leading from one end to the other. [Dr. Huxley's American Addresses, Lecture 2, p. 46] We come to cretaceous and no part of such series can be shown. So far as the present evidence goes, there is a break in the continuity of mammalian life in the Cretaceous period.
>
> I have also attempted to show that there was a break in the continuity of mammalian life in the Glacial epoch, which occured in the Pleistocene period. Now either of these breaks proves fatal to Dr. Darwin's hypothesis of evolution.[15]

The evolutionists speak of the "missing link" as though there were only a few parts of some mammoth puzzle that were missing. But contrary to the myth which has surrounded the evolutionist's more "public" utterances, all that they have of the puzzle are a few "links."

A second observation would be that the more fossil remains that the paleontologist finds, the more sure he is that fossil could not have been a direct ancestor to man. He finds too much specialization, too much that reveals the complexity of the specimen. It is interesting to note in this regard that a recent publication that has had quite wide appeal among the general public begins with *Philopithecus* and includes fourteen intermediary steps in the evolutionary process to modern man. All the steps are fully drawn, apparently to scale, complete with fur, skin, feet, legs, torsos, arms, and heads, But an examination of the actual fossil evidence for such could fit in a single coffin of a modern man! The less that is known, the more freedom the evolutionists apparently believe they have to speculate.

Third, as was noted by evolutionist Wilfred Le Gros Clark in apology of the eagerness of some of his colleagues, "Apart from other considerations, it is an interesting but not generally recognized fact that practically none of the genera and species of fossil hominids which have from time to time been created have any validity at all in zoological nomenclature. A newly named genus or species only becomes valid if it is accompanied by a formal diagnosis which clearly states in what respects the new type differs from another known genera or species, and it is perhaps rather surprising that paleo-anthropologists have not been in the habit of conforming to this important principle of taxonomy."[16] This could mean taxonomically, that the fossil evidence does not show more than variation within a species and not many separate species. For example, with the *Pithecanthropus* group there seems to be a high variability in the features, such as the relative development of the frontal region of the skull and the size of the jaws and teeth, and their cranial capacity actually ranged from 775cc to as much as 1,200 cc.[17] It has been found that Gall, the famous phrenologist; Anatole France, the French novelist; and Gambetta, the French statesman, each had the brain capacity of about 1,100 cc. If the skull of Anatole France and

of a *Pithecanthropus* were placed side by side, the
evolutionist would say that they were of similar creatures—
both sub-human. And, of course, they would be wrong.

A fourth observation is that all the "men" listed cannot be
members of man's linear ancestry. Evolutionist William
Howells, in his text *Mankind in the Making* explained,
"Obviously no living animal can be our ancestor, and this
point should never be forgotton."[18] What Howells was
saying in terms of man's ancestry is that no animals contemporaneous with each other could form a linear genealogy.
An example: Cro-Magnon Man is contemporaneous with
modern man. He cannot therefore be an ancestor to man. He
may have been a cousin. He may have degenerated from
modern man. But he can never be his lineal ancestor. And
any time we find two animals contemporaneous with each
other we may rule out one or the other or both from man's
genealogy.

A fifth observation would have to do with the fragmentary nature of the fossil evidence. Certainly the
evolutionists must feel apologetic. But Howells tells us that,
"It is not easy to become a fossil." He explains that,

> . . . If a creature dies, its bones may be promptly
> eaten and digested by hyenas or other animals.
> Otherwise they will lie on the ground and decay in a
> few years. They will not last much longer if they are
> buried in the earth. But if a carnivore drags the bones
> into a cave, like those in South Africa, where they
> become impregnated with hard and insoluble mineral
> salts, or if they fall into a swamp or a lake bed or
> some other place where they may last a long time
> until minerals have a chance to replace or join the
> material of the bone, then you have a fossil. All this
> counts against fossils of early man. For he was
> probably a rare animal in the first place and also, on
> the whole, too clever to fall into swamps or get eaten
> by a carnivore.
>
> That is only the beginning: a fossil may be
> destroyed again in the ground by a change in the soil
> chemistry. For example, the graves at that particular
> spot at Galley Hill actually contained no fossils other
> than the skeleton. [The Galley Hill Man] Any true

fossils have evidently been leached out by the acid in the ground, and so if the Galley Hill Man had been really old he would not have been there at all. Forests likewise have acid soils and hold little prospect for fossils, even though ancient stone tools may abound in them. Therefore, whole areas and whole periods, in which we know men were present, are devoid of human bones. We cannot tell what sort of men they were. There is nothing we can do about it but hope.[19]

And while that may satisfy Howells, it cannot explain why only skulls, jaws, and teeth were found. Certainly these parts are just as susceptible to all the forces of the elements, the acid soils, and the ravenous dogs that their other parts were. What happened to them? Paleontologists must shrug their shoulders and wag their heads, for they do not have the answers. When they tell us that the fossil evidence is the only sure foundation of evolution, that here is where evolutionists stand and fight their academic battles, we would expect more than a bunch of obscure, sometimes unidentifiable bones. That doesn't look like the kind of evidence even a high school freshman would take to his first debate meet.

How can a paleontologist take a rather common looking skull and catalogue it as a pre-historic man, a retarded ape, or a small female orangutan? What is their *modus operandi*? To understand how paleontologists work, let's take a hypothetical trip to Olduvi gorge in Tanganyika, South Africa. You have been digging all day when to your surprise you find a skull fragment. Eureka! There's not much of it; but you are certain that it is the "missing link." You know that it is a link in the evolutionary chain between the ape, ape-men, men-ape, and man. So this is what you are taught to do: First, look at the fossil fragment; second, you ask yourself "What does this look like—realizing of course that the end-result of whatever-it-is is modern man;" third, you look again toward the fossil and say, "Yes, this fossil confirms my observation—this is definitely an intermediary between ape and man." This method is actually used by paleontologists today. According to evolutionist, Le Gros Clark, paleontologists are taught,

> Trends of evolution *can be inferred* from a consideration of the end-results. Fossil evidence **may** confirm the fact of these trends and, if sufficiently complete, may demonstrate an actual evolutionary sequence in terms of successive types. It is useful to recognize this difference between an evolutionary trend and an actual linear, or ancestor-descendant, sequence, for it is only rarely that the fossil record of any taxonomic group is sufficiently abundant to permit the establishment of a true linear sequence. In the case of hominid evolution, the fossil record is still not adequate to allow firm conclusions regarding the entire linear sequence of the Hominidae which culminated in the emergence of Homo sapiens, but it is sufficiently adequate to demonstrate some of the main evolutionary trends which have occurred in human phylogenesis.[20]

It really is unbelievable that such circular reasoning is so beautifully camouflaged. But it is there: We can infer trends from the fossils, then the trends help to establish the order of the fossils! Fortunate for science there are some level-headed scientists and teachers. Not all scientists are evolutionists. Not all evolutionists accept everything from paleontology and not all paleontologists are evolutionists.

After considering the welter of jaws, teeth and partial skulls a sixth observation is that paleontologists have a difficult job trying to reconstruct the biological history of all living things. Relis B. Brown in his text *Biology*, observed,

> The piecing together of the evolution story is comparable to the reconstruction of an atom-bombed metropolitan telephone exchange by a child who has only seen a few telephone receivers. We know something about living plants and animals, and we have some fossil remnants to go on. Extensive study of the evidence available plus ingenious hypotheses, most of which cannot be adequately tested, have given us a sort of a trial schedule of the possible directions of evolution of living organisms.[21]

Zoologist Fleishmann said, "The study of palaeontology has not fulfilled the hopes that Darwin and his contemporaries placed in it."[22]

A final observation regarding the fossil evidence is that one of the most puzzling or most disturbing occurences for a palaeontologist is to find very complex fossil forms resting on very 'old' rocks such as granite; or finding simple forms of life contiguous to and underneath more complex forms. Walter F. Lammerts has reported that he has found "over 500 cases that attest to a reverse order, that is, simple forms of life resting on top of more advanced types."[23] With all the irregularities of the strata it is no wonder that geologists (palaeontology is a division of geology) find it difficult if not impossible to reconstruct the past.

It is fairly obvious that evolutionists who use palaeontology as a proof of man's evolution suffer from wishful thinking. The evidence they profess to find is only the mirroring of what they wish they had. Fortunately, we are not dependent upon the theory of evolution for our existence; and it is apparent that some evolutionists do not depend upon facts for theirs.

Chapter Eight

THE SCIENTIFIC METHOD AND EVOLUTION

In 1959, at the Darwinian Centennial celebration of the publication of *Origin of Species* at the University of Chicago Sir Julian Huxley declared to the world: "We all accept the fact of evolution. . . . The evolution of life is no longer a theory. It is a fact."[1] And since that time evolutionists have reaffirmed among themselves that Darwin's hypothesis, now dubbed law, would be the basis of their thinking.[2]

One must look a little askance at the proceedings of the centennial celebration; for a theory of science just isn't elevated to the status of law when there is so much contradictory evidence against it.

"Well," one may ask, "how does a theory become a law. Just what is the procedure, if what happened at the University of Chicago wasn't 'scientific?' " To answer, we must correctly begin our understanding of this "scientific method" by a brief definition of what "science" is.

Aristotle taught that science was "a sure and evident knowledge obtained from demonstrations."[3] Thomas Aquinias said that it was "the knowledge of things from their causes."[4] Here, then are two not dissimilar views: that science involved demonstrations, and then a knowledge of that thing is known from its cause. G. G. Simpson explained::

> "The proposed answers [concerning science] must, again by definition, be in **natural terms** and testable in some material way. On that basis, a definition of science as a whole would be: Science is an explora-

tion of the material universe that seeks natural,
orderly relationships among observed phenomena
and that is self-testing."[5]

Anthony Standon said that science is simply, "any
knowledge that is arrived at by the Scientific Method."[6] Our
task, then, comes to this: determining what the scientific
method is. But there have been volumes of writings about
what the scientific method is; so how do we define it?

Very simply the scientific method involves four steps.
First the scientist OBSERVES that which he is studying. He
describes, measures, calculates, diagrams, charts,
systematizes his thinking concerning a given phenomenon.
Second, he FORMULATES A HYPOTHESIS; that is to say,
he guesses concerning its probable occurence, cause, effects,
structure, shape, size, or anything else about which he
cannot directly observe. His third step, and this is crucial to
the entire Scientific Method, is to devise a method to
PROVE THE HYPOTHESIS WRONG! Now, most of us just
think that a scientists job is to try to demonstrate its
correctness. But that is not so—at least in the hypothetical,
scientific, detached world. Here we find, however, that
scientists are just ordinary people. Many of them have 8
a.m. to 5 p.m. jobs just like many other professional people
do. They are no more religious or irreligious than any other
group within a similar socio-economic level in society. And
they are no less susceptible to bias or prejudice or partiality
than any other professional man. Because they are not
endowed with omniscience and other qualities which many
times "press agents" bestow upon them, this third step is
particularly hard on them.

But its difficulty does not in any sense lessen its
necessity. As a matter of fact, the most sure way to
demonstrate bias or prejudice on the part of a scientist, or
for that matter upon any other person who has devised some
sort of hypothesis or other, is to ask him if he has conceived
of any way that his hypothesis could be demonstrated to be
incorrect. If he says, "Why, yes, as a matter of fact, I have
thought that if I just found evidence so and so, or if such
and such an experiment were to fail, then my hypothesis

would prove untenable," then you can rest assured that that man is up to this point, at least, using the Scientific Method, and would to that extent be considered a "scientist." But if he just looks at you blankly, or begins to get ruffled or squirmy, then take note. That fellow has not considered his hypothesis carefully enough yet, or is not really concerned with its proof. In that case he is trying to pawn off a philosophy and not a science. He should be called a philosopher, not a scientist.

The final step in the Scientific Method would be to PERFORM THE EXPERIMENT and then EVALUATE it. If after performing such-and-such an experiment, and the hypothesis is found inadequate, then discard it and start again with another one. If the hypothesis measured up reasonably well, then only slight revisions may be necessary.

When the scientist has "perfected" his hypothesis, ironed out all the wrinkles, he will most often publish it, thereby making it available to his colleagues for more wide-spread experimentation and criticism. If the hypothesis survives a great number of experiments, it is elevated to the position of theory. "A theory is simply a well-tested hypothesis, but there is no sharp dividing line," explains Standon. "Even the very best of theories may turn out to be wrong, for tomorrow an experiment may be done that flatly contradicts it."[7] But if it survives, if it is demonstrated to be true, if it succeeds with each experimentation, and more often than not after many years of careful qualitative analysis and thorough application of the Scientific Method, then the theory becomes a Law.

But being a law does not in any way mean that it is correct in an absolute sense. No, not at all. Any good physicist will tell you of the time—and that, not too far in the past—when it was completely agreed upon that there was no relation whatever between mass, matter and energy. But along came Einstein and others to demonstrate the invalidity of their theories and laws.

Euclidian geometry works wonders, at least in a practical sense, bqt it completely falls apart when applied to a curved surface. Being a law or theory of science only implies that no one yet has demonstrated the inadequacy of what it has implied.

We should not consider a "theory" as being without merit or importance. Indeed, science as we know it today would be non-existent without them. Some very useful theories have never been established as laws in any formal sense. Explaining the need for theories, Geologist Melvin A. Cook wrote,

> *Theory plays an important role in all arts and sciences (1) by providing a means for the unification and classification of available knowledge, and (2) by suggesting and prescribing the design of experimental studies that will broaden the scope of knowledge. Failure to accomplish either of these objectives necessitates moderations in the theory or substitution of an alternate one. For this reason the basic concepts are continually undergoing change in a healthy and forward-moving science. . . .*
>
> *One need not look far into science to discover it consists too generally of a maze of facts and theory so closely interwoven that even the most learned and honorable scientist (to say nothing of the intellectually dishonest one or the novice) may have difficulty distinguishing really between truth and theory. While this weakness of science is serious enough in fields which are not closely related to the primary purposes of mortality, in the fields more closely related, the difficulties of discerning fact and theory may well prove disastrous. This is particularly true as regards the development of spirituality in those who place science foremost.*[8]

Regarding the dangers involved in ascertaining the difference between fact and theory, it should be remembered that most theories are forever changing as newly discovered truth appears. And any time we accept as unchangeable truth the transitory theories of science we are skating on pretty thin ice; or as Scientist John A. Widtsoe observed, "Most theories are forever changing as new truth appears. That is the main reason why one cannot build firmly and finally on a theory and feel assured that he is on the safe road to truth."[9] Unfortunately, too many so called scientists do build on tentative theories, which, in turn, lessens the validity of their work.

It is this eagerness of some individuals to hold to even the most tentative of the hypotheses of science, claiming continually that "Science has done this and so," or "This is what science has proven." It is this sort of thing that caused George A. Zellers to write:

> Science has become a word to be conjured with, for which real scientists are not to be held responsible. There is nothing under the sun that may not be dubbed, or dub itself, science. Even the strolling fakir who sells a fake razor-strop paste may proclaim his wares the latest triumph of modern science, and by implication, if not explicitly, set himself forth as a scientist. Yet, as a matter of fact, his goods may be worthless, and he an ignoramus. . . .
>
> To scrutinize such science or scientists and to expose them must not be mistaken for an attack upon science, nor be understood as being in opposition to genuine scientists.[10]

Scientists and scholars pride themselves on their impartiality and sense of fair play. For the most part, they succeed admirably. But there have been not a few times when evolution has been defended, not in fairness and objectivity but in the spirit of battle.

After reading *Origin of Species* Huxley wrote to Darwin saying: "I trust you will not allow yourself to be in any way disgusted or annoyed by the considerable abuse and misrepresentation which, unless I greatly mistake, is in store for you. Depend upon it, you have earned the lasting gratitude of all thoughtful men. And as to the curs which will bark and yelp, you must recollect that some of your friends, at any rate, are endowed with an amount of combativeness which (though you have often and justly rebuked it) may stand you in good stead. *I am sharpening up my claws and beak in readiness.*"[11]

Geneticist Beatrice Bateson wrote of Huxley's militant spirit, and his ever-ready defense of evolution that swayed public sentiment to the hypothesis. She wrote, "To the world, scientific as well as lay, Huxley is chiefly famous as the champion of evolutionary doctrine, whose vigorous and skillful advocacy counted for so much on obtaining the favourable verdict of the public."[12]

Wilfred E. Le Gros Clark recently explained why there was so much unscientific controversy and emotional heat over the evolutionary hypothesis. He explained,

> Undoubtedly, one of the main factors responsible for the frequency with which polemics enter into controversies on matters of paleo-anthropology is a purely emotional one. It is a fact (which it were well to recognize) that it is extraordinarily difficult to view with complete objectivity the evidence for our own evolutionary origin, no doubt because the problem is such a very personal problem. Even scientists of today may not find it easy to clear their minds entirely of an emotional element when they come to consider the evidence in detail, and this emotional element is only too frequently betrayed by the phraseology with which disputants claim with equal insistence to be assessing the same evidence dispassionately.[13]

Bias and prejudice are difficult to eradicate from one's personality. But what of their opposite: open-mindedness? Is that a negative thing? Professor of History and Religion Hugh Nibley has answered "It is, unless it is a searching mind. An oyster has few prejudices—In the field of astronomy it has, we may safely say, absolutely none. Are we then to congratulate the oyster for its open-mindedness?"[14] He then quotes scientist J. B. S. Haldane who defined prejudice as "an opinion held without examining the evidence." Then said Nibley,

> Prejudice does not consist in having made up one's mind?in defending an opinion with fervor and determination—as too many liberals seem to think; it consists in forming an opinion before all the evidence has been considered. This means that freedom from prejudice whether in the field of science or any other field requires a tremendous lot of work—one cannot be unprejudiced without constant and laborious study of evidence; the open mind must be a searching mind. The person who claims allegiance to science in his thinking or who is an advocate of the open mind has let himself in for endless toil and trouble.[15]

But prejudice, forming an opinion or making up one's mind before all the facts are in, or interpreting the facts out of a faulty framework, have plagued man since he first began to think of his environment. Physicist Henry Eyring reports that there are two conflicting methods by which observable data are explained.

> The work of Rene Descartes and Sir Isaac Newton typifies [these] two methods.
>
> Descartes attempted to build up a universal system which would explain all the problems of nature from a unified philosophical point of view. He was quite willing to go far beyond what he could demonstrate experimentally, provided the point of view seemed rational to him. He thought everything, even problems of physiology, should be explained in terms of mechanisms. Most of his ideas on physics have been superseded by the painstaking method of careful observation. He fared better in the field of mathematics where he verified his results as he proceeded.
>
> Newton, on the other hand, was careful in his law of gravitation, in his mechanics and in his theories of optics to proceed slowly and to avoid going beyond what he could prove. Thus, imperfect measurements of others led to estimates of the moon's force of attractions to the earth one-sixth greater than was correctly predicted by Newton's theory. The result is that most of what Newton published still stands. Relativity and quantum mechanics simply extend Newtonian mechanics without superseding it in the realm for which it was developed.[16]

These two methods are with us today. One, in the form of true science; and the other, the all encompassing philosophy which could—if true—bind all nature into one tight package, is evolution. And if one doubts that evolution has become the almost universally accepted theorum from which all action springs, a look at some of the texts in the general field of science should prove instructive.

Biologist Sir Julian Huxley wrote:

> The concept of evolution was soon extended into other than biological fields. Inorganic subjects such

as the **life-histories of stars** and to the formation of
the **chemical** elements on the one hand, and on the
other hand subjects like **linguistics, social
anthropology,** and **comparative law** and **religion,**
began to be studied from an evolutionary angle, until
today we are enabled to see evolution as a universal
and **all-pervading process.**[17]

Geologist Carl O. Dunbar declared,

We now know, of course, that different kinds of
animals and plants have succeeded one another in
time because **life has continually evolved;** and
inasmuch as organic evolution is world-wide in its
operation, only rocks formed during the same age
could bear identical faunas.[18]

Professor of Astronomy Gerald S. Hawkins wrote:

We should expect to find **stars and galaxies in all
stages of evolution** as they form from existing
material and then decay. For stars this is certainly
the case. . . . There may be a similar evolutionary
process for galaxies, but at the moment we do not
have enough experimental evidence to give us the
clues to the evolutionary pattern.[19]

Psychiatrist Henry W. Brosin, chariman of the Department of Psychiatry at the University of Pittsburgh said,

It is appropriate for psychiatrists and other
students of mental disorders to pay homage to the
work of Charles Robert Darwin and the theory of
evolution, for without this work it is difficult to
imagine what the state of our discipline would be
like.[20]

Chairman of the Council of the British Sociological
Society Victor Branford wrote for the *Encyclopedia Britannica* the following:

The master idea, which animated alike the initiator
of sociology (i.e., August Comte) and his chief
continuator (Herbert Spencer), was that of evolution.
. . . Independently of the writings of both Comte and
Spencer, there proceeded during the 19th century,

> under the influence of the evolutionary concept, a thoroughgoing transformation of older studies like History, Law and Political Economy; and the creation of new ones like Anthropology, Social Psychology, Comparative Religion, Criminology, Social Geography. It is from these sources that have sprung the main body of writings, investigation, research, that today can properly be called sociology.[21]

C. L. Prosser correctly observed that "The origin of species has had more influence on Western culture than any other book of modern times. It was not only a great biological treatise, closely reasoned and revolutionary, but it carried significant implications for philosophy, religion, sociology, and history. Evolution is the greatest single unifying principle of all biology.[22]

Notice any relation between evolution and Descartes' philosophy?

Chapter Nine

EVOLUTION: A KIND OF FAITH

Paul told the Hebrews that faith was the substance of things hoped for, and the evidence of things not seen (Heb. 11:1). Evolution is something like that. The evolutionists maintain that there is evidence for their hypothesis, and yet a close examination of what they call evidence amounts to "things not seen."

Theodosius Dobzhansky testified, "The fact remains that among the present generation no informed person entertains any doubt of the validity of the evolution theory in the sense that evolution has occurred."[1] Arthur Thompson wrote, "We do not know any competent naturalist who has any hesitation in accepting the general doctrine. . . no one has any hesitation in regard to that fact."[2] H. H. Newman, author of *The Nature of the World and of Man,* taught, "Scientists the world over agree that the validity of the principle has been amply demonstrated. . . . Let us rest assured that the truth of evolution is demonstrated."[3]

But testimonials is about all that the searcher will find. Darwin testified that embryology is the key to his hypothesis; yet Sir Arthur Keith, Surgeon, and Albert Fleishmann, anatomist, both declare that there is no evidence of evolution in embryology. Anthropologist George Gaylord Simpson testifies that it was with the fossil record that evolution will defend its position, for here is the solid foundation of the theory; but Professor Louis T. More, says that the fossil record can demonstrate no such thing. Evolutionist William Berryman Scott surmised that morphology is the answer to the evolutionist's hopes; but Yale professor Carl O. Dunbar demonstrated that morphology is not what the evolutionists had claimed.

It is not speculative testimony that demonstrates the validity of a hypothesis. The Scientific Method demands evidence. But where is it? Where is all the proof that led Julian Huxley to announce that, "It takes an overwhelming prejudice to refuse to accept the facts, and anyone who is exposed to the evidence supporting evolution must recognize it as an historical fact."[4] The answer is simply this: There is no evidence. There is only a hope and a blind faith.

"The more one studies palaeontology," said Physicist Louis T. More, "the more certain one becomes that evolution is based on faith alone; exactly the same sort of faith which is necessary to have when one encounters the great mysteries of religion." He continued,

> The changes that are noted as time progresses show no orderly and no consecutive evolutionary chain and, above all, they give us no clue whatever as to the cause of variations. Evolutionists would have us believe that they have photographed the succession of fauna and flora, and have arranged them on a vast moving picture film. Its slow unrolling takes millions of years. A few pictures, mostly vague, defaced and tattered, occasionally attract our attention. Between these memorials of the past are enormous lengths of films containing no pictures at all. And we cannot tell whether these parts are blanks or whether the impression has faded from sight. Is the scenario a continuous changing show or is it a succession of static events? The evidence from palaeontology is for discontinuity; only by faith and imagination is there continuity of variation.[5]

One of the editors of the French Encyclopaedia, Paul Lemoine observed that the theory of evolution is being perpetuated by some scientists who do not themselves believe in its premises or conclusions. He told,

> It results from this expose that the theory of evolution is impossible. Moreover, in spite of appearance

no one no longer believes it [sic], and one says it,
without attaching any importance to it otherwise,
evolution in order to signify enchainment—or, more
evolved, less evolved in the sense of more perfectioned, less perfectioned, because it is conventional
language, admitted and almost obligatory in the
scientific world. **Evolution is a sort of dogma in
which the priests no longer believe but that they
maintain for their people.**

That—one must have the courage to say it in order
that men of future generations orient their research
in another way.[6]

Thornwell Jacobs wrote,

Master minds from all fields of discovery. . . are
united in their confession of faith which is embraced
in that superb generalization called 'evolution.'[7]

It is not an isolated case when an evolutionist proclaims
his dogma without grounds for his position. Discussing the
ease with which evolutionists promulgate their faith the
author of the book The Immense Journey, Dr. Loren Eiseley,
found,

A tendency for the beginning zoological textbook to
take the unwary reader by a hop, skip, and jump
from the little steaming pond or the beneficient
chemical crucible of the sea, into the lower world of
life with such sureness and rapidity that it is easy to
assume that there is no mystery about this matter at
all, or if there is, that it is a very little one.

This attitude has indeed been sharply criticized by
the distinguished British biologist Woodger, who
remarked some years ago: "Unstable organic
compounds and chlorophyll corpuscles do not persist
or come into existence in nature on their own account
at the present day, and consequently it is necessary
to postulate that conditions were once such that this
did happen although and in spite of the fact that our
knowledge of nature does not give us any warrant for
making such a supposition. . . . It is simple
dogmatism—asserting that what you want to believe
did in fact happen."[8]

If faith is the substance of things hoped for, then certainly evolution is a faith. For, as evolutionist William L. Strauss, Jr. explained,

I wish to emphasize that I am under no illusion that the theory of human ancestry which I favor at the present time, can in any way be regarded as proven. It is at best merely a working hypothesis whose final evaluation must be left to the future. What I am trying to point out is that, from what we now know, this interpretation appears to be distinctly more valid than the orthodox, anthropoid-ape theory. The ultimate verdict, if there can be a final verdict in such a matter, will rest upon paleontological evidence **at present lacking;** for with due respect to the **Australopithecinae,** the gap in the fossil record between man and the other primates remains very great indeed.

What I wish especially to stress is that the problem of man's ancestry is still a decidedly open one, in truth, a riddle. Hence it ill behooves us to accept any premature verdict as final and so to prejudice analysis and interpretation of whatever paleontological material may come to light as the orthodox theory has so often done and is still doing. One cannot assume that man is made-over anthropoid ape of any sort, for much of the available evidence is strongly against that assumption.[9]

If dogmatism is holding to a belief in spite of apparent contradiction; or without substantiation from facts; then evolution is a dogma. G. A. Kerkut reported that,

The serious undergraduate of the previous centuries was brought up on a theological diet from which he would learn to have faith and to quote authorities when he was in doubt. Intelligent understanding was the last thing required. The undergraduate of today is just as bad; he is still the same opinion-swallowing grub. He will gladly devour opinions and views that he does not properly understand in the hope that he may later regurgitate them during one of his examinations. Regardless of his subject, be it Engineering, Physics, English, or

Biology, he will have faith in theories that he only dimly follows and will call upon various authorities to support what he does not understand. In this he differs not one bit from the irrational theology student of the bygone age who would mumble his dogma and hurry through his studies in order to reach the peace and plenty of the comfortable living in the world outside. But what is worse, the present-day student **claims** to be different from his predecessor in that he thinks scientifically and despises dogma.[10]

Scientist do not like to be called "dogmatic" for it connotes apparent weakness in one's position. But what else could you call a position so weakly defended as that evolution is a fact?

Anthropologist J. M. Gillette considered evolution to be no more than dogma. He writes:

It is evident that anthropologists assume and profess to be evolutionists. They probably would resent the suggestion that they are dogmatists in the biological field, yet such appears to be the case regarding the subject under discussion. By their works ye shall know them, and it is by their work products in this particular field of evolution that they are now to be judged. Their genealogical trees purporting to show the evolution of man should satisfy theological fundamentalists who reject the idea of such evolution. In fact they are empty forms which consist of nothing but assumed roots, trunk, many limbs which grow in number through the years, and human twigs terminating the trunk which are supposed to connect with the assumed roots. Now a tree that is constituted wholly of limbs does not tell us much. Limbs below do not beget the limbs above them. They are not ancestral to them, only cousins to what is above them.

Biologists are, of course, confessedly evolutionists, but it is really remarkable how little evidence they admit in support of their position.[11]

If faith is the evidence of things not seen, then evolution is a faith. Biologist D'Arcy Wentworth Thompson describes the "evidence" in these words:

There is one last lesson which coordinate geometry helps us to learn; it is simple and easy, but very important indeed. In the study of evolution, and in all attempts to trace the descent of the animal kingdom, fourscore years' study of the **Origin of Species** has been an unlooked-for and disappointing result. It was hoped to begin with, and within my own recollection it was confidently believed, that the broad lines of descent, the relation of the main branches to one another and to the trunk of the tree, would soon be settled, and the lesser ramifications would be unravelled bit by bit and later on. But things have turned out otherwise. . . .

The larger and at first sight simpler questions remain unanswered; for eighty years' study of Darwin-ian evolution has not taught us how birds descend from reptiles, mammals from earlier quadrupeds, quadrupeds from fishes, nor vertebrates from the invertebrate stock. The invertebrates themselves involve the selfsame difficulties, so that we do not know the origin of the echinoderms, of the mulluscs, of the coelenterates, nor of one group of protozoa from another. The difficulty is not always quite the same. We may fail to find the actual links between the vertebrate groups, but yet their resemblance and their relationship, real though indefinable, are plain to see; there are gaps between the groups, but we can see, so to speak, across the gap. On the other hand, the breach between the vertebrate and invertebrate, worm and coilenterate, coilenterate and protozoan, is in each case of another order, and is so wide that we cannot see across the intervening gap at all.[12]

Evolutionist Sir James Gray says of Darwin's prize mechanism by which evolution is to have taken place,

We either have to accept natural selection as the only available guide to the mechanism of evolution, and be prepared to admit that it involves a considerable element of speculation, or feel in our bones that natural selection, operating on the random mutations, leaves too much to chance. . . . If we look on organic evolution as one of Nature's games of chance it seems just a little strange that she should have dealt quite so many winning hands. But, your guess is as good as mine.[13]

When one's guess is as good as another's in matters of this scientific inquiry, then such an admission would lead one to wonder just how plausible the entire scheme of evolution is. Certainly it would lead one to seriously question the purported evidence for evolution, for this single question recurs: If there are facts with which evolution can be demonstrated to be a scientific LAW, then why are the evolutionists seemingly doing all in their power to conceal them?

Says James H. McGregor in his text, *General Anthropology*, "Practically all enlightened people have come to accept the idea of man's origin by descent from lower animals, even though they may be quite ignorant of the evidence for it or the stages in the slow progression from simple beginnings to mankinds' present estate."[14] But it must also be concluded that no one is aware of the evidence of "the slow progression from simple beginnings to mankind's present estate." Certainly no one has thus far presented anything but wishful thinking, pie-in-the-sky theorizing, or a few faint and certainly indistinct mumblings concerning some extremely fragmented bones that have been found.

According to Geologist J. William Dawson,

The simplicity and completeness of the evolutionary theory entirely disappear when we consider the unproved assumptions on which it is based and its failure to connect with each other some of the most important facts in nature; that in short, it is not in any true sense a philosophy, but a mere arbitrary arrangement of facts in accordance with a number of unproved hypotheses. Such philosophies, falsely so-called, have existed ever since man began to reason on nature, and this last of all is one of the weakest and most pernicious of all. Let the reader

take up either Darwin's great book or Spencer's
Biology and merely ask, as he reads each paragraph,
What is here assumed and what is proved? and he
will find the fabric melt away like a vision. Spencer
often exaggerates or extenuates with reference to
facts and uses the art of the dialectician where
argument fails.[15]

And if any one doubts this all that need be done is to simply
read any text on the subject of physical, biological, cultural,
religious, or human evolution. You will be amazed at what
unfolds before you eyes! Error is made truth, fiction is made
fact, imagination is made evidence, and hypothesis is made
a scientific law!

William Jones long ago remarked that evolution was "a
metaphysical creed and nothing else; an emotional attitude
rather than a system of thought."[16] Tyndall observed: "To
the eyes of an onlooker their pace and method seem to be
like a steeple-chase. They are armed with a weapon always
sufficient if not always an arm of precision, 'the scientific
imagination.' They are impatient of that most wholesome
state, a Suspended Judgment."[17] Dawson taught,

It seems to indicate that the accumulated facts of
our age have gone altogether beyond its capacity for
generalization, and but for the vigor which one sees
everywhere, it might be taken as an indication that
the human mind has fallen into a state of senility and
in its dotage mistakes for science the imaginations
which are the dreams of its youth.[18]

Even one of the guiding prophets of the evolutionary faith,
who worked closely with Charles Darwin, T. H. Huxley,
was forced to admit that evolution was "an act of
philosophic faith," and that "there was no evidence that
anything of the sort had occured recently." Said Huxley's
son, "He discussed the rival theories of spontaneous generation and the universal derivation of life from precedent life,
and professed his belief as an act of philosophic faith, . . ."[19]

In a letter Huxley wrote to Charles Lyell, June 25, 1859, he
confessed, "I by no means suppose that the transmutation
hypothesis is proven or anything like it. But I view it as a
powerful instrument of research. Follow it out, and it will

lead us somewhere; while the other notion is like all the modifications of 'final causation,' a barren virgin."[20] He then let his eagerness for "what will be found in the future" *interfere* with "what is demonstrated to be proven today" *slip out. He said to Lyell:*."And I would very strongly urge *upon you that it is the logical development of Uniform-itarianism, and that its adoption would* harmonize the spirit of Paleontology with that of Physical Geology."[21] There was extremely little demonstrated in Huxley's day to warrant the conclusion that uniformitarianism would harmonize the spirit of paleontology with that of physical geology, anthropology, or biology. And the evidence today is no better.

Nibley's *G-2 Report No. 3* includes the statements of Bock, Jacobs, and Dobson, and Saunders who reveal further the difficulty with evidence in evolution.

> *K. E. Bock told, "It was long ago recognized that evolution was a dead horse—but there was nothing to take its place: 'This theoretical* **bankruptcy has forced us back into the evolutionist fold in spite of ourselves.'** *Since we must have some 'methodological framework within which we can speak generalization about cultures.' "*[22]

> *M. Jacobs observed, "Although details of most of these changes are* **inadequately evidenced** *by the fossil discoveries, the anthropologist uses as his* **frame of reference** *the concept of developmental levels. This serves as a means of classification in time.* **He presumes** *that* **subsequent discoveries** *will fit into one or another revision of an always tentative scheme of levels. This is the course of evolutionist thinking."*[23]

> *Dodson and Saunders stated, ". . . it may be well to recall the admonition of Hyman that 'the exact steps in the evolution of the various grades of intervertebrate structure are not and presumably* **never can be known.** *Statements about them are inferred from anatomical and embriological evidence and* **in no case should be regarded as established fact."**[24]

In view of the lack of evidence for the evolutionary hypotheses, Hassell has proposed "Articles of Faith" for the evolutionists—considering their failure to meet the requirements of science and the Scientific Method. Kerished slightly they are:

1. We believe that a lifeless, plantless ocean evolved out of itself aquatic plants; and then a marine vegetation, passing from its proper domain, became terrestrial; seaweeds thus transformed themselves into mosses, and mosses into ferns; and so like produced unlike.

2. We believe that cryptogamic vegetation, planned for itself floral organs, and altered its structure to suit such change. Regardless of there being no scientifically admissible evidence, we thus believe.

3. We believe acrogenic stems became endogenic, and some of these changed themselves into exogenic, and thus throughout the long vista of geological ages plants produced others not after their own kind, which thing, though contrary to experience, nevertheless did occur.

4. We believe that at some unknown period in the past the whole course of the vegetable world reversed itself, and from that time to this every plant has produced another after its own kind. Why persistency of species is now found to be the order of nature, while in the past transmutation pertained, cannot be determined; yet since the doctrine of Evolution requires that both be believed, it is to be accepted without questioning.

5. We believe that the first animals were evolved either out of non-living matter, or else from vegetable protoplasm. The primitive animals thus produced were destitute of any specialized contrivances for the performances of the functions of animal life,—respiration, circulation, assimilation; each was extemporised by the lump of jelly as occasion required.

6. We believe that as all animals were at first aquatic, but are now both aquatic and terrestrial, the latter were evolved out of the former; although there is no reason why such a thing should take place. But as the existence of land animals cannot be accounted for in any other way, it is to be believed, even though it is unsupported by any evidence.

7. As the invertebrated animals have their main masses of nervous matter ventrally disposed, and the vertebrates dorsally, by some unaccountable freak of nature the animal world was, once at least "turned upside down." It is difficult to say why, this should have taken place, or how it was accomplished; but inasmuch as the doctrine of Evolution requires that it did take place, that is enough,—therefore it is to be believed that it did occur.

8. We believe that every special organ in animals sprang into existence, as required, by the operation of the mystery of mysteries "natural selection," and so it came to pass that the oil-glands in the water birds were invented by a clever old goose who once suffered with rheumatic fever consequent upon repeated drenchings. After many failures, she hit upon this plan to prevent the mischief in the future.

9. We believe that birds were evolved out of reptiles, scales becoming feathers, fins becoming wings and feet; swim-bladders becoming lungs; a heartless creature extemporised a heart; two-chambered hearts became four chambered; and cold blood became hot. How, when, where, why, need not be known: suffice that it must have been so, because evolution requires it.

10. We believe that Class Mammalia being evolved out of reptiles or birds—it doesn't matter which—it came to pass, by some unaccountable act of the mystery of "natural selection," the form of the blood corpuscles were changed from oval to spherical, and the blood capillaries enlarged their capacity to suit the change. How this was accomplished it matters not. The unreasonableness of the whole affair makes it the more credible.

11. We claim that in the past, species were not fixed, and so it happened that one race of animals gave birth of another quite unlike itself; and so by the mystery of Evolution, a marsupialian was evolved into a ruminant, a ruminant into a rodent, a rodent into one of the quadrumana, and one of the quadrumana into one of the bimana. The unreasonableness of this is not to be questioned.

12. We believe that human speech and moral consciousness have been evolved as necessity occurred, and although

the highest forms of the quadrumana have never shown any tendency, during the human period, to advance towards a state of civilization, the very fact that they do not should be accepted as a proof that at one time they did. True, such a line of argument is illogical; but, then, if such changes did not take place Evolution cannot be true. It stands, therefore, that as Evolution must be true such changes did take place, notwithstanding their unreasonableness.

13. We believe that out of nonentities came potentialities; by the action of the non-living came life; by the motions of the inorganic were produced the organic; and by the commingling of the atoms of gross matter were produced thought, will, and conscience. Though all this is opposed to human reason and common sense, it matters not; it must be believed.[25]

Such is evolution: a faith, a dogma, a philosophy, but certainly not a science.

Chapter Ten

NATURAL SELECTION: DARWIN'S MECHANISM

The doctrine of Evolution teaches us that all living organisms are continually in a state of readjustment to their ever-changing environment. All life is ceaselessly struggling to survive. Only those organisms which change or adapt to their hostile environments are able to prepetuate their life's forces.

The evolutionists tell us that originally fish had fused backbones. But because they were able somehow to exert force on their spines, that is, break their backs thereby creating segmented vertebrae, they were able to survive.

Evolutionists teach that the earliest animals were without legs. According to the hypothesis, by some freak of nature a legless creature found on its body "slight escresences" or warts, which aided materially its progress as it wiggled along, and thus it acquired the habit of using them. This habit was transmitted to its posterity and they too, continued the habit until these warts lengthened and strengthened themselves by continued use, until they became legs.

Darwin's disciples teach that animals have survived only because sometime in the very dim past eyes were formed which gave the eyed creature an advantage over the blind one. They tell us that eyes were formed originally from some animal having pigment spots or freckles on the sides of its head, which, when turned to the sun, agreeably affected the animals so that it acquired the habit of turning that side of its head to the sun. Its posterity inherited the same habit and passed it on to still other generations. The pigment spot, the evolutionists swear with an oath, acquired sensitiveness by use; and in time a nerve developed which was the beginning of the eye.

We can be fortunate that in untold eons ago there was a great drought; for, according to the theory of evolution, in a time of drought some water animals, stranded by the receding waters, were required to adopt land manners and methods of living. Were it not for this imaginary drought, all of us would be living in the oceans instead of on the lands.

Another stroke of luck. The same drought that made aquatic animals terrestrial turned one terrestrial animal aquatic and into a whale! During the drought the land animal was required to seek food near the water's edge. Finally, with no use for legs they dropped off, and were replaced with fins and a flat tail. Darwin had inferred after seeing a bear swimming in a pool and catching insects with his wide-open mouth that terrestrial became aquatic.

Bless that drought! For what caused fish to become land animals, land mammals to become sea mammals, has also caused short-necked giraffes to grow longer necks. As the drought was just getting started, according to evolution, the herbage on the lower branches of the trees withered, leaving only that up higher to be eaten. Giraffes were required to stretch their necks; thereby after much tribulation and starvation their offspring were born with longer necks.

This recital of the evolution of legs, eyes, water mammals, and long-necked giraffes is true (according to the evolutionists). Only the scientific-sounding stilted language has been changed to facilitate learning. It is hardly necessary to reply to these fanciful speculations, but the following should be understood:

1. It is purely speculation. Not a single such change is known or has been observed. No evidence—not morphological, not embryological, not paleontoligical, not historical—can demonstrate these things.

2. "Acquired Characteristics" are not transmitted. Physical irritation or the use or disuse of a member of the body are not inherited. Only congenital characteristics in the fertilized egg cell are transmitted. [1]

"Natural Selection, as the prime evolutionary force in the evolution of human form," say Noel Korn and Fred

Thompson in their book *Human Evolution,* 'is the major principle enabling paleontologists and archaeologists to make sense out of the welter of bone fragments and broken artifacts that makes up their respective raw data."[2] They further tell us that Natural Selection "is the major agency of evolutionary change," but that it "does not merely act to redistribute genetic material in the gene pools of populations as a response to alterations in their ecological niches; it acts to maintain an equilibrium between the genetic potential of a population and its conditions of life."[3]

Tyndall taught that "Natural selection acts by the preserv-ation and accumulation of small inherited modifications, each profitable to the preserved being."[4] Wallace declared, "It is a fundamental doctrine of evolution, that all changes of form and structure, all increase in the size of an organ, or in its complexity, all greater specialization or physiological divisions of labour can only be brought about, inasmuch as it is for the good of the being so modified."[5]

But in spite of these testimonials concerning the importance of Natural Selection there are some glaring holes in the theory. In the first place the theory does not account for all the known facts of heredity. Clark and Mould observe that the "theory does not clearly explain why some variations are inherited and others are not. Many variations," they explain, "are so trivial that they could not possibly aid an organism in its struggle for existence.

Second, the theory does not explain how elongated warts, for example, could possibly evolve into highly complex, limbs. How blind evolution could ever produce nerve endings, blood vessels, specialized tissues, and specialized muscles is not explained by Natural Selection. And it appears that no evolutionist is concerned with its proof. Without Natural Selection evolution would be in shambles. "Let's not rock the boat," seems to be their attitude.

A third problem with Natural Selection is that there could be no possible immediate survival value in an elongated wart or the beginnings of an eye. "Evolution does not look ahead," we are told by evolutionist William Howells, "its eye fixed on the distant future. Instead, it is always trying to do its best with the business at hand. In doing this—in keeping some kind of animal well adapted to the life it is living, by one device or another—it sometimes stumbles into a new avenue leading off at a tangent, into a whole new earthly career."[6] If Natural Selection were as "mindless" and "purposeless" as evolutionist Mellersh indicated,[7] then the sheer luck by which life has been produced is the most unbelievable phenomenon in our history. But there is not a single fact to support the idea that there ever was a wart that turned into a leg, or a freckle that evolved into an eye, or that a broken bone became a flexible vertebrae.

Another problem with Natural Selection is what is called co-existence of types. What this means is that where there are two animals living side-by-side, one cannot have evolved from the other. The reason for this is simple. If the motivation to evolve in the first place was to survive, what would prompt an organism to evolve into something else if it could survive just as well without evolving? Nothing. Evolutionist Howells said, "Obviously no living animal can be our ancestor, and this point should never be forgotten."[8]

A fifth criticism of Darwin's contribution is that his theory explains nothing. Why do we find sheep that have obviously survived, that are able to provide their daily fare without evolving the long necks which were required by the giraffe for him to survive? The writer of an article entitled "Should We Burn Darwin," appearing in *Science Digest,* said,

> Perhaps the most significant single fact in last year's development of French scientific thought is that the . . . orthodox explanation of evolution has been badly shaken. Often criticized in the past, it has now come under such heavy fire that the way seems to be open, in France at least, to a new theory of the origin of species . . . .
>
> These are a few of the embarrassing questions asked today by the French rebels: If the giraffe with his eight-foot neck is the product of natural selection and an example of the survival of the fittest, what about the sheep with its neck no longer than a few inches? Aren't giraffes and sheep very close cousins, almost brethren in the animal kingdom . . .? But then can there live side by side two cousins, each of them

*fitter than the other, one because its neck is longer,
the other because its neck is shorter?*

*And talking of sheep, what about their horns?
According to the classical school they started
growing freakishly, and then, as they proved an asset
in the sheep's struggle for life, nature went on selecting the horned animals and eliminating the hornless
ones. But did it really? There are at least as many
hornless sheep as those with horns. Which of them
are fitter? . . .*

*Out of 120,000 fertilized eggs of the green frog only
two individuals survive. Are we to conclude that
these two frogs out of 120,000 were selected by
nature because they were the fittest ones; or
rather . . . that natural selection is nothing but blind
mortality which selects nothing at all?*[9]

Sixth, if Natural Selection created new species, then
another problem arises. No one has successfully arrived at a
definition of species that squares with both the evidence
and the theory. Evolutionist William Bateson said it was a
group of organisms with marked characteristics in common
and freely interbreeding.[10] An associate with Bateson in the
British Association for the Advancement of Science,
Professor Poulton called it "an interbreeding community."[11]
Contrarily, Theodosius Dobzhansky asked,

*"Can we define species? If we require that a defini-
tion of species always enables a biologist to decide
whether the forms he observes are distinct species or
merely races of one species, then nobody has
invented a definition. It may, moreover, be doubted
that such a definition will ever be invented.
Strangely enough, such a 'success' would come close
to overthrowing the theory of evolution. . . ."*[12]

He then told why a definition of species—as we have it
defined by Bateson and Poulton—would be fatal to the
theory of evolution. He said,

*Genetically effective interbreeding is absent
between species. Contemporaneous species do not
exchange genes, or do so but rarely. There is, for
example, no living species with which man could
interbreed. Although the horse and donkey species*

are hybridized on a large scale to produce mules, mules are wholly, or almost wholly sterile, so that no gene interchange results. . . ."[13]

Summarizing the evidence of the creation of new species California Institute of Technology Professor T. H. Morgan declared:

Within the period of human history we do not find a single instance of the transformation of one species into another one. It may be claimed then that the theory of descent is lacking in the most essential feature that it takes to place it on a scientific basis.[14]

McNair Wilson, M.D., editor of Oxford Medical Publications told that an

Increase of knowledge about biology has tended to emphasize the extreme rigidity of type and, more and more, to discount the idea of transmutation from one type to another—the essential basis of Darwinism. The classic aphorism, 'when a mule breeds,' ought to serve as a warning against the easy acceptance of a theory which is as full of ogres, mermaids and centaurs as any fairy tale.[15]

It is ironic, isn't it, that the men who know the most about science seem to accept theories of evolution the least?

A seventh problem with Natural Selection was mentioned by Charles Darwin himself. He noted.

Why, if species have descended from other species by fine gradations, do we not everywhere see innumerable transitional forms? Why is not all nature in confusion, instead of the species being, as we see them, well defined?

But, as by this theory innumerable transitional forms must have existed, why do we not find them embedded in countless numbers in the crust of the earth?

Geological research . . . does not yield the infinitely many fine gradations between past and present species required.[16]

The doctrine of Natural Selection requires that there be an immense number of "failures" in nature. Since the whole impelling purpose of evolution is to survive, and since the natural process of selecting characteristics in each organism which over many generations allow it to survive would undoubtedly cause failures, then where are the fossil remains of these failures? Simply stated, there are none.

Evolutionist A. S. Romer admitted, "Unfortunately there is in general little evidence on this point in the fossil record, for intermediate evolutionary forms representative of this phenomenon are extremely rare. 'Links' are missing just where we most fervently desire them, and it is all too probable that many 'links' will continue to be missing."[17]

D'Arcy Thompson in his evolutionary interpretation *On Growth and Form* wrote:

> Eighty years' study of Darwinian evolution has not taught us how birds descend from reptiles, mammals from earlier quadrupeds, quadrupeds from fishes, nor vertebrates from the invertebrate stock. The invertebrates themselves involve the selfsame difficulties, . . . the breach between vertebrate and invertebrate, worm and coelenterate, coelenterate and protozoan, . . . is so wide that we cannot see across the intervening gap at all . . . .
>
> We cross a boundary every time we pass from family to family, or group to group. . . .
>
> A 'principle of discontinuity,' then, is inherent in all our classifications, . . . to seek for stepping stones across the gaps between is to seek in vain for ever.[18]

Again, evolutionist Dobzhansky observed,

> If we assemble as many individuals living at a given time as we can, we notice at once that the observed variation does not form any kind of continuous distribution. Instead, a multitude of separate, discrete, distributions are found. The living world is not a single array in which any two variants are connected by unbroken series of intergrades, but an array of more or less distinctly separate arrays, intermediates between which are absent or at least rare.[19]

And another evolutionist was forced by the evidence to admit,

> *Unfortunately, the fossil record which would enable us to trace the emergence of the apes is still hopelessly incomplete, we do not know either when or where distinctively apelike animals first began to diverge from monkey stock.*[20]

The editors of *Life Magazine* in the publication of *The World We Live In*, state that "For at least three-quarters of the book of ages engraved in the earth's crust the pages are blank."[21] And the *New York Times* wrote, "Even today surprisingly little is known of man's own family tree . . . . There are still enormous gaps."[22]

The idea of "Missing Links" is a myth. They have not found them, and, as has been noted, the chance that they will ever be found is not even being considered by reputable paleontologists.

Again we must come to the realization that evolution is a hypothesis without scientific basis, the forming of fact into fancy; the creation of a tenuous philosophy into a Golden Calf for both worship and amusement by the evolutionists.

Chapter Eleven

The Magic of Life

Perhaps the earliest account of man's search for an answer to the secret of life was the ancient prophet of Israel, Moses. He had told the Children of Israel that "the blood is the life," and that they were not to "eat the life with the flesh."[1] But whether Moses actually thought that the vitality of animal life was a quality of the blood, or that he gave only a superficial justification for the ancient peoples not eating the blood with the flesh, cannot be known from the Biblical record. One thing is certain, though, and that is that even until the advent of the very most recent scientific discoveries concerning life and its qualities, it was thought that the blood was the life.

A London anatomist, Dr. John Hunter, for example, as late as the nineteenth century had "proved" to his satisfaction that the blood was equivalent with life, and that it provided the vitality of all animal life. Biblical scholar Adam Clarke summarized Dr. Hunter's "proofs" by proof-texting Leviticus 17:11 saying:

> 1. That the blood unites living parts in some circumstances as certainly as the yet recent juices of the branch of one tree unite with that of another; and he thinks that if either of these fluids were dead matter, they would act as **stimuli,** and no union would take place in the animal or vegetable kingdom; and he shows that in the nature of things there is not a more intimate connection between **life** and a **solid** than between **life** and a **fluid.**
> 2. He shows that the blood becomes **vascular,** like other living parts of the body; and he demonstrated this by a preparation in which **vessels** were clearly

seen to arise from what had been a *coagulum of blood; for those vessels opened into the stream of the circulating blood, which was in contiguity with this coagulating mass.*

3. *He proves that if blood be taken from the arm in the most intense cold that the human body can suffer, it will raise the thermometer to the same height as blood taken in the most sultry heat. This is a very powerful argument for the* **vitality** *of the blood, as it is well known that living bodies alone have the power of resisting great degrees of heat and cold, and of maintaining in almost every situation while in health that temperature which we distinguish by the name of* **animal heat.**

4. *He proves that blood is capable of being acted upon by a stimulus, as it coagulates on exposure to the air, as certainly as the cavities of the abdomen and thorax become inflamed from the same cause. The more the blood is alive, i.e., the more the animal is in health, the sooner the blood coagulates on exposure; and the more it has lost of the living principle, as in cases of violent inflammation, the less sensible it is to the stimulus produced by being exposed, and coagulates more slowly.*

5. *He proves that the blood preserves life in different parts of the body. When the* **nerves** *going to any part are* **tied** *or* **cut,** *the part becomes paralytic, and loses all power of motion, but it does not mortify. But let the* **artery** *be cut, and then the part dies and* **mortification** *ensues. It must therefore be the vital principle of the* **blood** *that keeps the part* **alive;** *nor does it appear that this fact can be accounted for on any other principle. . . .*[2]

If religious faith must be built on this sort of "scientific" experimentation, then heaven help the religious! Fortunately for us all, true science of today is more sophisticated than it was a hundred years ago. But Clarke's statement does (though obviously tedious) illustrate a point. It shows the foolishness, first of all, of those who would proclaim the conclusions of "science" at any one time as "proving" something conclusively. There might be an overwhelming abundance of evidence for a conclusion. It may seem in every respect correct. But the moment we stamp anything as "proved by science" then we are letting ourselves in for a great deal of trouble; for there is nothing so constant in the pursuit of scientific truth as change. Second, it illustrates the ridiculous position in which religion is placed almost every time that it sides with the theories of the day.

False theories, inaccurate dogmas, man-made "revelations," and a host of other characteristics of false religion all must give way to truth, regardless of its source. The apostle Paul told the ancient saints at Thessalonica to "Prove all things; hold fast that which is good,"[3] but it is obvious that many of the teachers of religion have not followed the advice. It is proof-texting like the current example that leads many who seek the truth to reject both the Bible and religion.

There is evidence that the life processes were being observed as early as Aristotle who broke open hens' eggs and studied the growth of the embryos. But as to the origins of life, Aristotle and his contemporaries knew nothing. In 1653 an English surgeon named William Harvey erroneously wrote, "An egg can no more be made without the assistance of the Cocke and Henne, than the fruit can be made without the trees aid."[4] Biologist Ernest Havemann traced the history of his science, saying,

> Throughout the Middle ages and much of the Renaissance, scientists considered each individual as preformed from the very moment of conception. Only about 300 years ago, with the discovery of the microscope, did they begin to discern the facts. The male sperm was seen and described in detail; the egg was carefully examined and seen to have its own structures, not those of an adult. Today's scientists can detect the finest structures of egg and sperm, analyze their chemical content and study the embryo's growth from the first moments of life. All that has been learned, however, only increases the sense of awe with which man beholds the beginnings of life.[5]

From the first crude experiments with the microtome and microscope to the present with highly sophisticated electron microscopes, high-speed photographic equipment, and the learning of the past 300 years, scientists are intensifying their search into the mysteries of life. Biology has come a long way since its early days. Biologists, biochemists, and anatomists and scores of other specially trained scientists are very confident of the future. Said biochemist Kornberg, a pioneer in working with DNA, "Within ten years, it will be possible, to modify genes to produce specific biologic changes in animals and human beings. The discovery might enable scientists to create artificial viruses which would attack and kill cancers."[6] And while the successes in biology may warrant such an enthusiastic display of confidence, science must concern itself with present accomplishments and present problems. It is a far too easy thing to prophesy of future accomplishments. It could be comparable to the freshman in college who after having successfully passed a survey course and feeling no small amount of pride, boasts of his success without really understanding that wisdom is made up of more than a single classroom experience.

"The cell," according to biologist John Paul, "can be defined as the smallest organized unit of living form which is capable of prolonged independent existence and replacement of its own substance in a suitable environment."[7] In talking about life, then, the cell is perhaps the single most important unit. It is with the study of the cell that any study of life must begin.

It has been estimated that there are about between 60 to 100 trillion cells in the adult human being. Biologist Irving Adler observed that "the cells of living things are usually too small to be seen with the naked eye. The average cell," he says, "is so tiny that a line of 250 of them, arranged end to end, would be only one inch long."[8] End to end these 100 trillion cells would stretch 6.3 million miles, or two hundred times around the earth at the equator.

Biologist Ashley Montagu, speaking of the size and complexity of specialized cells—the sperm cell and the egg cell—noted,

> At the present moment [1959], there are about 2,750,000,000 human beings alive on the face of this earth. Allowing one sperm and one egg to each of them as being responsible both for their existence and their genetic heredity, we have a total of 5,500,-000,000 germ cells involved, a number which could

be contained in about two and a half quart milk bottles.[9]

He continues by saying that, "The sperm cells would occupy the space of less than an aspirin tablet. In fact, the chromosomes, the actual bearers of the hereditary genes, within the cells of this huge number would occupy the space of less than half an aspirin tablet."[10]

How a person can believe that all this economy and complexity in the world of life would have spontaneously and gradually evolved from non-living substances is beyond reason. After considerable research, scientist Jay M. Savage came to the conclusion that "life is a manifestation of much greater complexity than any non-living system [and] is underscored when we stop to realize that even the simplest cell is composed of thousands of different kinds of molecules operating together in a co-ordinated fashion."[11]

Nucleic acid, the genetic material of the cell, is composed of protein, but there are thousands of different kinds of proteins. "Protein comes in forms as various as silk, fingernails, skin, homones, enzymes, and viruses. Gelatin, egg albumin, casein and insulin are pure protein," wrote Stanley and Valens in their book, *Viruses and the Nature of Life.* They also tell that "some 50,000 different kinds of protein account for nearly half the dry weight of the human body."[12] These 50,000 kinds of protein are "manufactured" from just 20 kinds of amino acids.

The complexity of a protein can be illustrated by showing chemical notation of a single molecule of water as H_2O being composed of two atoms of hydrogen and one atom of oxygen. Casein, a relatively simple protein looks like this: $C_{708}H_{1130}O_{224}N_{180}S_4P_4$. Or, in other words, a single molecule of the protein casein is composed of 708 atoms of carbon, 1,130 atoms of hydrogen, 224 atoms of oxygen, 180 atoms of nitrogen, 4 atoms of sulphur, and 4 atoms of phosphorus. But it is not enough to simply combine the proper number of atoms of each element as a cook would prepare a cake batter. It takes agents, or reagents, to combine each of the elements in their proper sequence and prepare each chemical for combination at a specific energy level. And yet evolutionists tell us that all this happened as a result of blind chance. Writer Bruce Barton, reflecting on these claims, said,

*When you can dump a load of bricks on a corner lot
and let me watch them arrange themselves into a
house . . . it will be easier for me to believe that all
those thousands of worlds could have been created,
balanced, and set in order in their various orbits, all
without a designing intelligence at all.*[13]

And Herbert Ross, author of *Understanding Evolution*,
wrote of the complexity and importance of the cell,

*From the original intake of raw materials from the
environment to the duplication of the last molecule in
the mature cell, these tiny machines must lead from
one to the other, the last dependent on the first and
the first on the last.*

*A failure of even one could conceivably bring the
whole chain to a halt, resulting in the death of the
organism. For one of the essential facts about life is
that it must keep going to stay alive.*[14]

If we speak of the cell as a machine, what do we know of
its internal parts? Until quite recently virtually nothing was
known of all but the more prominent parts: the nucleus,
cytoplasm, the proteins, the chromosomes, the genes. But
now biochemists, physicists, and even mathematicians are
turning their attention to the study of DNA, deoxy-
ribonucleic acid. Evolutionist Rutherford Platt explained a
few facts about it, saying,

*Your personal DNA is peppered throughout your
body in about 60 thousand billion specks—the
average number of living cells in a human adult. . . .*

*Surprisingly, the DNA molecule has a basically
simple form. It consists of two intertwined, tape-like
coils of lined-up atoms connected by cross pieces at
regular intervals—like a spiral staircase. . . . There is
logic in the long slender form of DNA: this gives it a
capacity, like magnetic recording tape, to store the
vast amount of data needed in a lifetime.*

*The DNA tapes themselves are of sugar and
phosphate; the crosspieces of the spiral staircase are
nitrogen compounds. . . . Their varying sequence on
the DNA tapes directs the events which make bodies
grow—much as the tiny variations on magnetic tapes*

produce, according to their order, the sounds of the music. . . .

Dr. Beadle says that if we were to put the coded DNA instructions of a single human cell into English, they would fill a 1000 volume encyclopedia.

All the while that DNA sits in the nucleus giving orders that will spur growth, digestion, heartbeat, thinking and feeling, it is following its built-in plan which it has carried down the corridors of time. It makes no alterations in that plan, unless they are imposed by radiations or accidents from outside the cell.[15]

We might think of the human body as a total system with each of the 60,000,000,000,000 cells comprising its structure as sub-systems to the complete body. Then within the cell is a sub-system of chromosomes, within which exist sub-systems of genes, and then a DNA sub-system within the genes. And just how many sub-systems will be discovered below DNA is anyone's guess.

To date, scientists have found no difference in chemical composition of the DNA in man or any other form of life. According to Rutherford Platt,

These DNA specks have a similar chemical composition, are about the same size, and look very much like those in your dog, or in a housefly, a bread mold or blade of grass. Yet somehow the specks are coded to make every living thing different from every other living thing. They make dogs different from fish or birds, bread mold from apple trees, elephants from mosquitoes.[16]

So, even with all that is known about the DNA molecule, there is a great deal yet to learn. Some scientists, however, were not content to wait until something more substantial is known. Almost immediately after the first news that something exciting was being done, that something was being learned about the life processes, there were those who proclaimed that science had opened wide the doors on the mysteries of life, that the artificial creation of life was soon to be accomplished—if it had not in fact already occurred! One such group of men, Drs. Kornberg, Goulian and Sinsheimer, working with viruses, claimed that they had created an artificial virus. When asked whether they had in fact created life, Dr. Kornberg replied, "There is no one accepted definition of the word, 'life.' " But added, ". . . With reservations I've mentioned it would be fair to say a viral DNA is a simple or primitive form of life."[17] Dr. Goulian said, "Different people mean different things by life. . . . If you grant a virus is alive or that naked DNA is alive, then this was a creation of life."[18] But what, in fact, had they done? Had they created life out of non-life? Scientist Paul Kroll explained,

> Dr. Kornberg put a naturally occurring DNA of a virus into a test tube. This common virus, known as PHi X-174 attacks intestinal bacteria. Note this carefully. The natural viral DNA in the tube would serve as blueprint or mold from which the artificial DNA would be made.
>
> Next, molecules of adenine, guanine, cytosine and thymine were added to the brew. These naturally occurring molecules are the basic units of DNA.
>
> Also, some E. coli DNA polymerase was also mixed in. This enzyme is crucial for guiding the copying procedure. This enzyme is obtained from living bacteria. Finally, another enzyme, ligase, was added to finish the duplication job.
>
> The contents were then centrifuged to separate the artificial batter DNA from the natural mold DNA. The artificial DNA was able to infect host cells and reproduce itself.
>
> Stop and think about this.
>
> The original DNA of the virus already existed; the chemical building blocks of the DNA already existed; the enzymes already existed! The scientists had the mold. They merely reorganized matter (an accomplishment to be sure!) using an already existing mold.
>
> But proof of evolution? Not at all!
>
> "Life" in a test tube? Ridiculous! Life evolving from nonliving? Doubly ridiculous.[19]

Why would the spontaneous creation of a virus from non-life be ridiculous? Why could not spontaneous generation have occurred in the dim past as the evolutionists proclaim?

the answer may have been given by an expert in viruses, Wolfhard Weidel. In his book entitled, *Viruses*, Weidel tells, "Even if some spontaneous act of generation had brought a virus particle forth from some sort of primordial slime, that particle would have remained lonely and forgotten without the simultaneous presence of living cells."[20] In other words, in order for viruses to have existed spontaneously, then living cells must have also spontaneously arisen from non-living matter within a reasonable period of time, and within reasonable proximity, too.

Scientist Hans Gaffron asked a searching question regarding the definition of life. He said,

> Matter can practically always be defined in terms of physics, chemistry and biochemistry. This certainly is not enough to define life.
>
> We might ask: If we ingest food, at what moment does the food become living? Of course it never does. One could follow a particle of assimilated food, no matter how complex, and wherever one finds it in the living organism, it is dead. It is the process in which it takes part that defines it, and not the matter of which it is composed. . . so the essence of life is found in the process of living and not in any constituents of living cells.[21]

Life cannot be defined in terms of DNA, proteins, or any other chemical composition. Accordingly, DNA cannot possess the vitality of life, which the ancients thought was possessed by the blood. Scientists are at an impasse for they cannot by scientific means explain or understand the magic of life.

But there is a place where a person can find the answer. The scriptures teach that "the spirit and the body are the soul of man."[22] Or, that it is the spirit, the creator of which is God, our Heavenly Father, that is the vitality of the physical body.

The body is as it were a glove, lifeless and without mobility. But when the spirit enters it, as a hand the glove, the metabolic processes of the body operate. What Science has created is not life but only delayed or prolonged chemical activity. On a very simple yet illustrative level, if one were to put some baking soda into a glass of vinegar the chemical action would begin immediately to neutralize the acid and base, resolving them to a passive chemical salt. Or, if the experimenter were to prepare an iodide "clock" which indicated the presense of the chemical iodine in a delayed action by turning a blue color, he would not be creating life—though the apparent reaction of the chemicals he combined could be delayed quite some time. Chemical action and reaction cannot define nor can it account for life.

Evolutionists have offered no evidence for their claims—that life was created spontaneously, yet most of them assume it to be true. So far their claim has amounted to scientific-sounding faith and nothing more.

Chapter Twelve
IN THE BEGINNING: SPONTANEOUS GENERATION

Virtually all evolutionists speak of their theory as an on-going process. Most of them believe it to be one of progression instead of just change, and many say that it is irreversible. But hardly anyone discusses the beginning of the evolution of life, and this is indeed strange. It is like saying that one accepts the corollaries of a theorem without understanding the premise. When this becomes our posture we let ourselves in for endless troubles. The absurdity of accepting the theory of evolution of life without an understanding of how they claim life was organized is immediately evident.

The Bible begins its account of the creation of life on this earth by saying "In the beginning God . . ."[1] But the evolutionists—for the most part—choose to replace God as the designer and executor of this planet with the theory of blind, directionless, self-perpetuating evolution. Instead of having God create or organize life on this planet they say that it began without assistance, that it was generated spontaneously.

Ernst Haeckel taught that life originated out of non-living substance, that life was evolved "by a process purely mechanical, from purely inorganic carbon, combinations, that very complex nitrogenised carbon compound which we call plasson, or 'primitive slime' and which is the oldest material substance in which vital activities are embodied."[2] Fortunately, other scientists are more cautious than zealot Haeckel. E. J. Gardner, in his text *Organic Evolution and the*

Bible, taught,

> *A type of spontaneous generation may have taken place in the remote past (a billion years or more ago) from which the forms presently living on the earth may have descended. . . .*
>
> *The possibility of the appropriate elements, energy and suitable environment coming together by chance seems remote, indeed, but in tremendously long periods of time the "impossible" becomes inevitable.*[3]

The editors of the Reader's Digest *Great World Atlas* simply assume the presence of life:

> *For its first billion and a half years the Earth was probably without life. The Earth's earliest crust may have been volcanic; the most ancient rocks have dominantly greenish colors, an indication of old lavas and their products. The early atmosphere was probably without oxygen, consisting, it is thought, of hydrogen, water vapor, methane and ammonia. An atmosphere rich in oxygen must have evolked at least three billion years ago; in rocks of that age we find fossils of primitive plants which used carbon dioxide and liberated oxygen. Sediments were washed into the ocean, shifting the weight of the crust, producing violent heat changes and creating numerous mountain systems. Widespread ice ages alternated with warmer times.*
>
> *Seaweeds and probably bacteria were the only forms of plant life.*[4]

Most evolutionists seem reluctant to formally commit themselves to the discussion of the spontaneous beginning of life, but merely assume its correctness and proceed with their own subjective verification of the theories of evolution.

But what of spontaneous generation? What is the evidence for it? Biologists Mark A. Hall and Milton S. Lesser explained how the scientific method was used to ascertain the incorrectness of the hypothesis of spontaneous generation. They reported,

> *Francesco Redi, an Italian physician, was the first (about 1688) to carry out controlled experiments that disproved the belief that maggots arose from*

decaying fish, snakes, and meat. . . . Redi proved that
maggots and flies arise from living parents, not from
dead matter.

  Lazzaro Spallanzani, an Italian priest (about 1780),
sealed numerous vegetable juices in glass flasks and
then boiled them. After allowing these materials to
cool and stand for a number of days, Spallanzani
could not observe any organisms. Even microscopic
examination did not reveal them. Spallanzani
concluded that nothing developed in the juices
because boiling killed any living organisms that
might have been present. Consequently, there were
no living organisms to give rise to new ones.

  Louis Pasteur, a French scientist (about 1860),
conclusively demonstrated that microorganisms,
which are present everywhere, get into organic
matter, which serves as their food. After feeding and
growing, the microorganisms reproduce and thereby
give rise to many others like themselves. If flasks
containing foo l are sealed and sterilized, . . . even
after many months, no microorganisms appear.[5]

Pasteur's claims were not immediately recognized,
however.

In 1870 a Dr. Bastian published his account of the
experiments which he performed. He had prepared infusions
of hay and turnips and placed them in glass tubes. He then
boiled the contents and sealed the ends of the still-steaming
tubes. After a while he examined the contents under a
microscope and in some of the tubes he found forms of
animal life! Just how life had spontaneously developed from
ostensibly non-living matter Bastian never tells; but his
experiment has been duplicated several times since, and
except when the glass tubes or flasks had been broken, or
the seal been destroyed, no life has been observed. Joseph
Hassell described perhaps the most dramatic disproof of
Bastian's spontaneous generation of life. He explained,

  In 1879 Dr. Tyndall performed a number of
experiments with a view of further testing the ques-
tion. He procured sixty flasks, in which he placed
infusions of beef, mutton, turnips, and cucumber. All
these infusions of beef, mutton, turnips, and

cucumber. All these infusions were boiled for a certain length of time, and while boiling the necks of the flasks were sealed. The Doctor now carefully packed up and removed them to his house at Ben-Alp in Switzerland, at an elevation of 7,000 feet above the sea. When the box was opened fifty-four of the infusions were found to be clear, and six were muddy. On close examination it was discovered that the flasks containing the muddy infusions were damaged, and, as a consequence, the air had entered. In these various forms of life were found to exist.

The fifty-four remaining flasks were now exposed for three weeks to the sun's rays by day, and to the warmth of a room by night; at the end of the time they were as clear as at the commencement. Four of the flasks were now damaged, and the fifty remaining were divided into two sets. Twenty-seven were carried up to a ledge of the Alps 10,000 feet above the sea. The ends of the flasks were now broken, and the whole were allowed to remain for a period of three weeks exposed to the wind which was blowing across the snow-capped peaks of the Oberland. At the end of three weeks the infusions were found as clear as they were before the exposure, and when submitted to microscopic investigation there were no traces of animal life.

The other twenty-three flasks were taken to a hay-loft in the rear of the Doctor's house; the necks were broken off, and the infusions allowed to remain for three weeks in direct communication with the air. At the end of the time the infusions were found to be muddy, and when submitted to microscopic investigation were found to be rich in animal life.

When the Doctor returned to London he performed a number of experiments under similar conditions, and in every case with similar results.

When speaking of these experiments, and supposing they had been investigated by a careful observer, he says, "Such faithful scrutiny fully carried out would infallibly lead him to the conclusion that, as in all other cases, so in this, the evidence in favour of spontaneous generation, crumbles in the grasp of the competent inquirer."[6]

The idea of spontaneous generation as with the idea of evolution did not, of course, originate with Charles Darwin. The ancient Greek Anaximander taught that men had evolved from fish, and Empedocles believed that animals had been derived from plants. It was also commonly taught that insects, fishes, the higher animals, and likely man were all originated out of mud or slime or some other inorganic substance.[7] It was also taught that crocodiles were generated from the slime of the Nile, bees and flies from decomposed flesh.[8] Of the theories of spontaneous generation Joseph Fielding Smith, in his comprehensive work, *Man, His Origin and Destiny*, explained,

> *This belief continued down to the middle of the nineteenth century especially in relation to insects and bacteria and lower forms of life, and it was the scientists who followed the ideas of Darwin, Wallace and other who were the keenest in the search to discover if this apparent notion were true. When careful research was made the whole theory exploded; and it was these scientists who were forced to admit that such a thing as spontaneous generation is not true. . . .*
>
> *Notwithstanding the great discovery of Pasteur, Darwin and his followers were not retarded in their search to find the beginning of life and to prove that all things have developed from spontaneous life. This question has never been answered successfully other than the account in the scriptures: If spontaneous generation cannot be created now, how could it be possible several million or billion years ago? Conditions, according to the teachings of science are more favorable now than they possibly could have been in the far distant past. To get a beginning these advocates must **assume** some starting point, notwithstanding there is no evidence that will support it. All evidence points to the contrary.[9]*

Even if life could be spontaneously generated in the laboratory, (which experiment has never been demonstrated in the history of the scientific method) the theories of evolution and spontaneous generation would still remain unproved. Success in a laboratory where a directing force—

the scientist—orders the elements, nudges the agents and reagents, prods the process along as best as he can, is not the blind direction-less accident which according to the evolutionists caused life to originally form itself. There can be no super-force, no supernatural explanation, no Master Intellect or Intelligence. Evolutionists are programmed to reject any supernatural explanation when a natural, physical one is even remotely possible—even if the possibility is only theoretical. G. G. Simpson explained, "In science one should never accept a metaphysical [supernatural] explanation for a physical explanation as possible, or indeed conceivable."[10]

Fortunately, a person need not reject God in order to call himself a scientist[11] any more than a person must accept the hypothesis of evolution to be considered a scholar. The following story regarding the faith of Sir Isaac Newton as he taught his atheistic friend of absurdity of spontaneous generation is instructive:

> One day, as Newton sat reading in his study with his mechanism on a large table near him, his infidel friend stepped in. Scientist that he was, he recognized at a glance what was before him. Stepping up to it, he slowly turned the crank, and with undisguised admiration watched the heavenly bodies all move in their relative speed in their orbits. Standing off a few feet he exclaimed, "My! What an exquisite thing this is! Who made it?" Without looking up from his book, Newton answered, "Nobody!" Quickly turning to Newton, the infidel said, "Evidently you did not understand my question. I asked who made this?" Looking up now, Newton solemnly assured him that nobody made it, but that the aggregation of matter so much admired had just happened to assume the form it was in. But the astonished infidel replied with some heat, "You must think I am a fool! Of course somebody made it, and he is a genius, and I'd like to know who he is."
>
> Laying his book aside, Newton arose and laid a hand on his friend's shoulder. "This thing is but a puny imitation of a most grander system whose laws

you know, and I am not able to convince you that this mere toy is without a designer and maker; yet you profess to believe that the great original from which the design is taken has come into being without either designer or maker! Now tell me by what sort of reasoning do you reach such an incongruous conclusion?"[12]

Far more complex than cranking a simple machine is the human eye, the brain, the auditory canal, the movement of the hand, the restorative powers of an individual cell, the nervous system, and thousands and thousands of other highly complex organic systems, including even the "simplest" of cells. We think of the complexity of the modern computers with their highly sophisticated electro-magnetic circuitry, with their virtually unerrancy computation, or the extensive electronic circuitry of a color television camera. But these man-made systems are puny when compared to the living systems of organic matter. One writer declared that "The cell is as complicated as New York City."[13] Evolutionist Loren Eiseley wrote of the complexity of the "simple" cell, saying, "To grasp in detail the physico-chemical organization of the simplest cell is far beyond our capacity."[14]

Zoologist Sir James Gray, Cambridge University teaches:

A bacterium is far more complex than any inanimate system known to man. There is not a laboratory in the world which can compete with the biochemical activity of the smallest living oganisms.[15]

Evolutionist John T. Bonner tells,

The cell is really such an astoundingly clever unit that when we think of it from the point of view of evolution it seems easier to imagine a single cell evolving into complex animals and plants than it does to imagine a group of chemical substances evolving into a cell. It is very likely that the first step was more difficult. . . . The study of early evolution really amounts to educated guesswork.[16]

And as example of how really complex, yet how efficient they are, we know this of the ability of cells of green plants to produce by photosynthesis:

> The largest single manufacturing process in the world takes place in one of the smallest units of life—the cells of green plants. The manufacturing process is . . . photosynthesis. Each year this process accounts for the transformation of 100 billion tons of the inorganic element carbon into organic forms that support life.
>
> By contrast, all the big blast furnaces of the world make only a half-billion tons of steel in the same time.[17]

The more that man develops skills that allow him to probe deeper into the mysteries of organic matter, the more he comes to the awareness that life is by no means simple. It is no wonder Edwin Conklin once remarked: "The probability of life originating from accident is comparable to the probability of the unabridged dictionary resulting from an explosion in a printing shop."[18]

Eisley told,

> Intensified effort revealed that even the supposedly simple amoeba was a complex, self-operating chemical factory. The notion that he was a simple blob, the discovery of whose chemical composition would enable us instantly to set the life process in operation, turned out to be at best, a monstrous caricature of the truth.
>
> With the failure of these many efforts science was left in the somewhat embarrassing position of having to postulate theories of living origins which it could not demonstrate. After having chided the theologian for his reliance on myth and miracle, science found itself in the unenviable position of having to create a mythology of its own: namely, the assumption that what, after long effort, could not be proved to take place today had, in truth, taken place in the primeval past.[19]

Bruce Barton pondered the ridiculous claims that life was created from non-life through the action of mindless evolution. He retorted: "When you can dump a load of bricks on a corner lot and let me watch them arrange themselves into a house—when you can empty a handful of wheels and springs and screws on my desk, and let me see them gather themselves into a watch—it will be easier for me to believe that all these thousands of worlds would have been created, balanced, and set in order in their various orbits, all with a designing intelligence at all. Moreover if there is no intelligence in the universe, then the universe created something greater than itself—for it created you and me."[20]

Biologist Frank B. Salisbury compared the spontaneous creation of protein molecules to the chance printing of the sentence, "This above all, to thine own self be true." By building a computer with the volume of one liter that could make a trillion trials per second, then stacking computers two kilometers deep over the entire surface of the earth, then stacking them onto 10×10^{20} planets (also to the depth of two kilometers) and having them work independently for four billion years the sentence would appear only once.[21]

Even evolutionist Thomas Henry Huxley found it difficult to accept the theory of spontaneous generation strictly from the evidence currently available. He admitted:

> Looking back through the prodigious vistas of time, I find no record of the beginning of life, and, therefore, I am devoid of any means of forming a conclusion as to the conditions of its appearance. Belief, in the scientific sense of the word, is a serious matter and requires strong foundations. To say, therefore, in the admitted absence of evidence, that I have any belief as to the mode in which life forms have originated, would be using words in the wrong sense.[22]

But he also said,

> If the hypothesis of evolution be true, living matter must have originated from non-living matter for, by that hypothesis, the conditions on the globe were at one time such that living matter could not have existed, life being entirely incompatible with the gaseous state.[23]

Biologist John Elliot Howard has written:

> If spontaneous generation is not true, if life can only proceed from life the whole doctrine of evolution fails at the very commencement. It is a very obvious and oft-repeated truth that no chain can be stronger than its weakest link, and the chain of reasoning

above referred to is **entirely wanting** in the first link. It hangs upon nothing! It has no answer to the inquiry, "Whence is the origin of life?" and the speaker is driven in his perplexity to adopt the most unscientific of all assumptions for the solution of the enigma, the suggestion of the impossible, as follows: — "It might be held that the conditions affecting the combination of the primary elements of matter into organic forms may at one time have been different from those which now prevail, and that under these different conditions **abiogenesis** may have been possible, and may have operated to lay the foundations or organic life in the simple forms in which it first appeared,— a state of things which can only be vaguely surmised, but in regard to which no exact information can be obtained."

Science is founded on the observation of fact, but evolutionism on the hypothesis that the reverse of all known facts may have been at some time true; the whole conditions affecting the combination of the primary elements of matter are rearranged to suit the theory. The quiet **assumption** that "organic life first appeared in simpler forms" is to be noted, and then the **candid** admission that this can only be vaguely surmised, and "no exact information can be obtained."

The whole passage is so complete a specimen of evolutionist argument, that I have not hesitated to present it entire. It is proverbially true that a man convinced against his will remains of the same opinion; and this, evidently, is the case with the Doctor, who first tell us that **abiogenesis** is impossible, then that it **must** have existed, and then that what we want now to complete the proof is exact information **how** it existed. I extract from a recently-published work by Mr. Darwin, a specimen of the kind of reasoning objected to. In speaking of the varieties of Primula, he says: —"We may freely admit that **Primula veris, vulgaris,** and **elatior,** as well as all other species of the genus, are descended from a common primordial form, yet, from the facts

*above given, we must conclude that these three forms
are now as fixed in character as are many others,
which are universally ranked as true species. Consequently they have as good a right to receive distinct
specific names as have, for instance, the ass, quagga,
and zebra."*

*It is always the same—***facts** *on the one side,* **theory**
on the other. On the **ipse dixit** *of Darwin we may
"freely admit" that of which no proof can be given,
and which is the direct reverse of all present
experience! Ouch is the faith that Darwin looks for
(and not in vain) from his followers![24]*

On June 25, 1859, T. H. Huxley wrote to Charles Lyell,
whose evolutionary theories of geology are with us still
asking: "How much evidence would you require to believe
that there was a time when stones fell upwards, or granite
made itself by a spontaneous rearrangement of the elemen-
tary particles of clay and sand? And yet," he said, "the
difficulties in the way of these beliefs are as nothing
compared to those which you would have to overcome in
believing that complex organic beings made themselves (for
that is what creation comes to in scientific language) out of
inorganic matter."[25]

Other scientists, in addition to Huxley, have found
difficulty in accepting spontaneous generation as answer to
the origin of life. Scientist Joseph LeConte wrote:

*If life did once arise spontaneously from any lower
forces, physical or chemical, by natural processes,
the conditions necessary for so extraordinary a
change could hardly be expected to occur* **but once** *in
the history of the earth. Yet they are now not only
unreproductible, but unimaginable.[26]*

Biologist Lorande Woodruff told,

*We thus reach the general conclusion that, so far as
observation and experimentation are concerned, no
form of life exists today except from pre-existing
life.[27]*

Physicist L. T. More wrote,

> To talk of the evolution of thought from sea slime to amoeba, and from amoeba to a self-conscious thinking man, means nothing; it is the easy solution of the thoughtless mind.[28]

Biologist Edmund Wilson maintained,

> As early as 1855 Virchow positively maintained the universality of cell division, contending that every cell is the offspring of a pre-existing parent cell. Today this conclusion rests on a foundation so firm that we are justified in regarding it as a universal law of development. The study of the cell has on the whole seemed to widen the enormous gap that separates the lower forms of life from the inorganic world.[29]

Medical doctor McNair Williams explained,

> Modern medicine and surgery are founded on the truth enunciated by Pasteur, that life proceeds only from life, and only from life of the same kind and type.[30]

Spontaneous generation is not only in violation of all that is learned by observation and experimentation, it is also in violation of a LAW of physics—the Second Law of Thermodynamics which states very simply that in a closed system—like our universe—the amount of "entropy," that is, the amount of DIS-organization in the system, will tend to increase. Or stated another way, the amount of chaos will increase. It can be seen imediately that the theory of evolution, which claims that all matter is increasing in order and complexity, wherein the amount of entropy would be decreasing, is in direct contradiction to this law of physics. Or as stated by the review of a recent book by R. Schubert-Soldern:

> "All molecules result from an electro-chemical tendency to neutralisation. They are therefore expressions of tendencies toward stability." Unhappily for materialists, however, life is characteristically unstable, and "it is incredible that

a complex of substances, all tending towards a state of stability, would produce the permanent chemical instability which is characteristic of animate matter." Thus it is inconceivable that an organic compound should ever be formed in the absence of life: "No condition of inorganic matter is even thinkable in which carbon, oxygen and hydrogen could combine to form a sugar rather than water and carbon dioxide."[31]

The hypothesis of evolution has again been weighed in the balance of scientific evidence and has been found wanting. As a science, evolution is frustrated, but as an imaginary faith formed like a Golden Calf it has real possibilities.

Chapter Thirteen

MUTATIONS: CHANGING FOR THE BETTER?

The word "mutation" comes from the Latin *mutars*, which means "to change" or "to alter." And when evolutionists use the word they mean "changes in the genes and variants of the gene structure," said geneticist Theodosius Dobzhansky.[1] These "changes in the genes"— mutations—"are the raw materials of evolution."[2] Evolutionist Amram Scheinfeld explained that,

> *It is through the rare instances of favorable mutations, of innumerable kinds and in countless numbers, occurring successively over very extended periods, that the whole process of evolution may now be explained.*[3]

It is pretty important, then, that a person understand something about mutations, their causes, and their effects; because without them evolution would not take place.

Though Charles Darwin never used the word "mutation" he did allow for that kind of minute, gradual change. He said, "Slight individual differences. . . . suffice for the work, and are probably the sole differences which are effective in the production of new species." Then he said that "if it could be demonstrated that any complex organ existed, which could not possibly have been formed by numerous successive slight modifications, my theory would absolutely break down."[4]

It is postulated by the evolutionists, for example, that animals, in the course of their evolution, gradually developed eyes and found them to be of tremendous survival value? Though they were rudimentary, unformed, and useless; and the animals were still blind. Darwin himself said of the eye and its complexity,

> "To suppose that they eye with all its inimitable contrivances for adjusting the focus to different distances, for admitting different amounts of light, and for the correction of spherical and chromatic aberration, could have been formed by natural selection seems, I freely confess, absurd in the highest degress."[5]

But the absurd becomes a way of life for the evolutionist. And so the eye was created by blind chance, they say. But the fossil evidence tells something quite contrary to Darwin's hypothesis. For example the Jellyfish *Medusa* which has existed for unnumbered ages has survived not with the perfect eye of the present, but "rudimentary" eyes. Yet man has never found a single particle of evidence that the rudimentary eye of the *Medusa* has ever been more than what it is today.

The trilobite possesses compound eyes, much as do the present-day insects. The evolutionists cannot say that the compound eye contributed to the demise of the trilobite, for insects have survived. Nor can they say that the ancient trilobite compound eye is of less utility, more primitive, more rudimentary than other eyes—they are just different, and a difference in structure does not imply evolution.

The spiders of today have a number of simple eyes, but are their fossil ancestors different than they? No, not at all. The eyes of the fossil spiders are identical to those of their modern offspring.

The fossil ants are identical to the same variety of modern ants.

Evolutionists hasten to the rescue of the fossil record by telling of modern fish which, they say, have forgotten how to use their eyes, or that they have evolved only rudimentary eyes. But we must draw their imaginations back to the statement by evolutionist George Gaylord Simpson, who said, "The most direct sort of evidence on the truth of evolution must, after all, be provided by the fossil record."[6] We must ask of such fish, "What fossil evidence do we have that they ever had eyes, if we must assume that they have fallen into disuse," or if the evolutionists tell us that the evolution of their eyes is not complete, that they have only eyes in their rudimental state, "What fossil evidence is there that that species of fish ever were without even the rudimental eyes which they now possess?"

We must conclude that if the paleontological evidence is "the most direct sort of evidence on the truth of evolution" as Simpson maintained, then evolution is built on sand and not rock. So we must turn from ancient fossils to the study of modern mutations.

The little fruit fly *Drosophila Melanogaster* has over the years been a favorite of the geneticists. Twenty successive generations of *Drosphila* can be reared in a single year, the equivalent of 25,000 years of man. And in 1927 it was discovered that by exposing the little fly to X-rays that the rate of reproduction can be increased some fifteen thousand percent! During the past forty years of experiments, the equivalent of which in human years being 150.8 million years, geneticists should have some remarkable evidence to provide evolutionists. They have found, for example, that the fly can develop red spots on its wings, no wings, different pigmentation, doubling of already existing members of the body, or deletions of parts of the body. But they have never produced anything other than the fruit fly. In other words, they have succeeded in demonstrating that mutations of *Drosophila* produces only variations, and never other species.

An evidence of evolution recently submitted to the world for investigation is the fact that certain moth larvae were fed on industrial smog, or Melanogen, which is believed to be either manganous sulfate or lead nitrate. Many of the fully developed moths produced darkened wings. The phenomonon is called "Industrial Melanism." But with this moth, as with the fruit fly, we find variation within the species, adaptation to its environment, but never a new species created.

The lack of proof of evolution by mutation has led evolutionist Jean Rostand to querry,

> Is it really certain, then, as the new-Darwinists maintain, that the problem of evolution is . . . a

settled matter . . . ? I, personally, do not think so,
and, along with a good many others, I must insist on
raising some banal objections to the doctrine of neo-
Darwinism. . . .

The mutations which we know and which are
considered responsible for the creation of the living
world are, in general, either organic deprivations,
deficiencies (loss of pigment, loss of an appendage),
or the doubling of the pre-existing organs. In any
case, they never produce anything really new or
original in the organic scheme, nothing which one
might consider the basis for a new organ or the
priming for a new function. . . .

No, decidedly, I cannot make myself think that
these "slips" of heredity have been able even with the
cooperation of natural selection, even with the
advantage of the immense periods of time in which
evolution works on life, to build the entire world,
with its structural prodigality and refinements, its
astounding "adaptations," . . . I cannot persuade
myself to think that the eye, the ear, the human brain
have been formed in this way; . . . I discern nothing
that gives me the right to conceive the profound
structural alterations, the fantastic metamorphosis
that we have to imagine in evolutionary history when
we think of the transition from invertebrates to
vertebrates, from fish to batrachians, from
batrachians to reptiles, from reptiles to mammals.[7]

The rate of mutation is extremely low. Said Geneticists
Bruce Wallace and Theodosius Dobzhansky, "Mutational
changes in any one gene are rare events. This is a different
way of saying that, ordinarily, the genes reproduce
themselves accurately."[8] Geneticist C. H. Waddinton
explained that "It happens rarely, perhaps once in a million
animals or once in a million lifetimes."[9] Another evolutionist
stated that "most genes mutate only once in 100,000
generations or more. Researchers estimate that a human
gene may remain stable for 2.5 million years."[10] And
zoologist Douglas Dewar reports that the mutation rate "is
of the order of 10^{-5} to 10^{-6} per gene per generation."[11]

In addition to the infrequency of mutations, another factor
to consider is that virtually all mutations are detrimental to

the organism. H. J. Muller, who received the Nobel Prize in 1946 for his work with *Drosphila melangoaster* told, "Most mutations are bad, in fact good ones are so rare that we may consider them all as bad."[12] Dobzhansky admits, "A majority of mutations, both those arising in laboratories and those stored in natural populations, produce deteriorations of the viability, hereditary diseases, and monstrosities. Such changes, it would seem, can hardly serve as evolutionary building blocks."[13] One scientist writes, "Of the many mutants detected in the laboratory, all are either recessives or "semi-dominants," and the majority cause harmful physiological effects. Hardly any have ever been observed which could possibly be beneficial to an organism under wild conditions."[14] And another said, "Most mutations are harmful. Some make it impossible for the cells in which they occur to develop and grow."[15]

Dr. W. E. Lammerts, horticulturalist, described his experimentation with roses:

> *My own work on neutron radiation of roses described a technique by which we can induce 50 radiated buds of Queen Elizabeth, more mutations than could hitherto be found in a lifetime of searching among several million rose plants grown annually from non-radiated buds. Without exception, all mutations induced were found to be defective or weaker than Queen Elizabeth. . . . Biologically they would hardly compete because of reduced vigor and partial sterility.*[16]

Muller reported that "in more than 99 per cent of cases the mutation of a gene produces some kind of harmful effect, some disturbance of function."[17]

A critical examination of the conclusions offered by competent geneticists allows for perhaps a one per cent chance that evolution occurs—if it does at all—by mutation. And while it cannot satisfy the searching inquirer, it is nevertheless true (at least philosophically (that as time approaches infinity even an infinitesimally finite probability may occur.

Even if the mathematical probability is conceded one still must answer the question of fossil evidence. The answer is of course obvious. There is no fossil evidence.

The assumption that the mutation of a single gene at a time is, according to evolutionist Waddington, like this:

> This is really the theory that if you start with any fourteen lines of coherent English and change it one letter at a time, keeping only those things that still make sense, you will eventually finish up with one of the sonnets of Shakespeare. . . . It strikes me as a lunatic sort of logic, and I think we should be able to do better.[18]

The evidence of evolution lies deep within the soul of the evolutionist, for it is there that his pulse for religious faith and furvor is found. Evolution is not science in the sense that its premises can be observed, documented, calculated, or tested. It is fashioned out of "negative evidence," and the imaginations of its formulators. Dr. A. Labbe, professor at the School of Medicine at Nantes, France, said of genetics,

> Genetics, which is consecrated to the study of heredity, has become a kind of religion, dogmatic, mystical, intolerant, which has its temples, its priests, its believers, its councils, and which aims at converting all the biologists in the world. . . . Genetics ends inevitably in a more or less complete negation of evolution: at the most it can conceive of fortuitous variations. . . . We do not want this genetics which hampers us.[19]

Douglas Dewar, whose work of many years in zoology has won him worldwide recognition, has summarized five reasons why the evidence of mutations does not satisfy the requirements of the theory of evolution. He lists:

> 1. The experimental work of geneticists and of practical breeders shows that species are very stable and resistant to attempts to transform them, despite the phenomenon of variation. . . .
> 2. The experimental work of geneticists seems to show that the effects of use and disuse are not inherited, nor are characters acquired by an individual during its life-time. . . .
> 3. The vast majority of mutations are the reverse of benefical; indeed a large percentage are lethal, i.e.,

they lead to the early death of the animal in which they occur. . . .

4. Another fact, which in my view is most unfavorable to the evolution theory, and which writers on genetics are apt to slur over, is the large number of genes which co-operate to produce quite trivial features. For example, as Stern admits (Genetics, Palaeontology and Evolution—1946): "No less than 30 genes co-operate in forming the actual colour of the eye of the adult **Drosophila**. From this it follows that if each gene operates in connection with only one character, the number of genes possessed by **Drosophila** is quite inadequate for the realization of all its characters. Therefore geneticists have to believe that most, if not all, genes affect a number of characters. As Stern puts it : "The conclusion follows, therefore, that in general there is no simple one-to-one relation of gene to character, or of character to gene. Development of organization, character and organism must accordingly be envisaged as consequences or products derived from multidimensional networks of genic innteractions." . . .

As a mutation seems to involve the dislocation or disturbance of at least one of the atoms in one of the molecules of the gene affected, the resulting mutation is likely to affect all the organs or features on which that gene acts, and the odds must be enormous against this effect being favourable on all or most of these organs. . . .

5. Of the facts brought to light by the geneticists and cytologists one of the most unfavourable to evolutionism is that the chromosomes of the simplest organism appear to be as complicated as those of the highest animals. "The chromosomes of some Protozoa," writes R. Goldschmidt (**The Material Basis of Evolution**—1940, p. 6) "look uncomfortably like those of the highest animals."[20]

That mutations occur is not in question, but the presumption that mutations are the mechanism by which natural selection or any other process of selection directs evolution, or that mutations have evolved the nearly one million species of life on earth from a single ancestor is clearly suspect. And until geneticists can come up with something more conclusive than what they have discovered, it would be extremely foolish to even infer that evolution can be demonstrated to have occurred by mutation. Or as scientist Hugh Miller noted,

> *The relative rarity of these aberrant or mutant changes, together with their usually maladaptive and more often than not lethal effects upon development, does not incline us to assign to them an important role in the maintenance of group-adaptability. . . . It should be observed that the great importance currently attached to gene-mutations as a factor in evolutionary history is in part the result of erroneous expectations initially aroused by their discovery.*[21]

The Golden Calf of Evolution appears a bit weather-beaten. And genetics is doing nothing to brighten it up.

Chapter Fourteen

CREATION AT THE CAMBRIAN?

One inevitable conclusion from reading both the scriptural and the geologic accounts of creation[1] is that the creation seems to have taken place with great suddenness. And even in spite of the fact that the Hebrew word for "day," yom, can mean an indefinite space of time, in addition to meaning the hours from sunrise to sunset, and a twenty-four hour period,[2] both the scriptures and scientific investigations teach that the processes of creation were at one point in time operable, and then suddenly in-operable.

Surely the position that the earth and its life were created magically by the cryptic words or the secret potion of an eternal wizzard must be discarded as unfounded in both science and religion. By "suddenness" of creation is meant a period relatively short, when compared to the supposed millions or billions of years of creation which evolutionists are pleased to report.

Paleontologists, for the most part, are unanimous in their conclusion that life appeared suddenly at what they have labeled the Cambrian period. Or in other words, the fossil record shows that life suddenly appeared at what experts have said was about 600 million years ago.

George Gaylord Simpson, Harvard University, taught,

> It remains true, as every paleontologist knows, that most new species, genera, and families and that nearly all new categories above the level of families appear in the record **suddenly** and are not led up to by known, gradual, completely continuous transitional sequences.[3]

Charles Darwin similarly observed that the geologic record (earlier than what is now called Cambrian) was blank. He remarked,

> Why, if species have descended from other species by fine gradations, do we not everywhere see innumerable transitional forms? Why is not all nature in confusion, instead of the species being, as we see them, well defined?
>
> But, as by this theory innumerable transitional forms must have existed, why do we not find them embedded in countless numbers in the crust of the earth?
>
> Geological research . . . does not yield the infinitely many fine gradations between past and present species required.[4]

Darwin further admitted,

> If we confine our attention to any one formation, it becomes much more difficult to understand why we do not therein find closely graduated varieties between the allied species.[5]

Botanist Heribert Nilsson, testifies,

> If we look at the peculair main groups of the fossil flora, it is quite striking that at definite intervals of geological time they are all **at once** and **quite suddenly** there, and, moreover, in full bloom in all their manifold forms. And it is quite as surprising that after a time which is to be measured not only in millions, but in tens of millions of years, they disappear equally **suddenly**. Furthermore, at the end of their existence they do not change into forms which are transitional towards the main types of the next period: such are entirely lacking.

Evolutionist Lecomte du Nouy confessed,

> Each one of these intermediaries seems to have appeared "suddenly," and it has not yet been possible, because of the lack of fossils, to reconstitute the passage between these intermediaries. . . . The continuity we surmise may never be established by facts.[6]

A recent article in *Scientific American* clarified the situation regarding the obviously sudden appearance of life in the geological record. It reported, in part,

> Both the sudden appearance and the remarkable composition of the animal life characteristic of Cambrian times are sometimes explained away or overlooked by biologists. Yet recent paleontological research has made the puzzle of this sudden proliferation of living organisms increasingly difficult for anyone to evade. . . .
>
> These animals were neither primitive nor generalized in anatomy: they were complex organisms that clearly belonged to the various distinct phyla, or major groups of animals, now classified as metazoan. In fact, they are now known to include representatives of nearly every major phylum that possessed skeletal structures capable of fossilization; . . .
>
> Yet before the Lower Cambrian there was scarcely a trace of them. The appearance of the Lower Cambrian fauna . . . can reasonably be called a "sudden" event.
>
> One can no longer dismiss this event by assuming that all Pre-Cambrian rocks have been too greatly altered by time to allow the fossils ancestral to the Cambrian metazoans to be preserved. . . . Even if all the Pre-Cambrian ancestors of the Cambrian metazoans were similarly soft-bodied and therefore rarely preserved, far more abundant traces of their activities should have been found in the Pre-Cambrian strata than has proved to be the case. Neither can the general failure to find Pre-Cambrian animal fossils be charged to any lack of trying.[7]

An article in *Natural History*, confirmed,

> From the beginning of the Cambrian up through the rest of the geological sequence, we have an abundant representation of animal life at every stage; even in Lower Cambrian formations, marine invertebrates are numerous and varied. Below this, there are vast thicknesses of sediments in which the progenitors of

*the Cambrian forms should be expected. But we do
not find them; these older beds are almost barren of
evidence of life, and the general picture could
reasonably be said to be consistent with the idea of
special creation at the beginning of Cambrian times.*

*"To the question why we do not find rich
fossiliferous deposits belonging to these assumed
earliest periods prior to the Cambrian system," said
Darwin, "I can give no satisfactory answer." Nor can
we today.*[8]

Six hundred million years ago, as the geologists calculate
time, life appeared suddenly from non-life, every phylum, or
great groups, of animals then existed. There is an indication
that even vertebrates existed then.[9] And an article
appearing in the New York *Times* a short time ago told that
there existed the same division between plant and animal
kingdoms as they appear today. It reported, "The chief
puzzle in the record of life's history on earth . . . [is] the
sudden appearance, some 600 million years ago, of most
basic divisions of the plant and animal kingdoms. There is
virtually no record on how these divisions came about. Thus
the entire first part of evolutionary history is missing."[10]

Darwin's theory of evolution presupposed that creation is
of infinite duration, that it has had no beginning, and that it
will never end. But Darwin himself was faced with the
stubborn fact that the geological record was not in harmony
with his theory; or said another way, Darwin's theory was
not in harmony with the evidence of geology. Darwin wrote:

*There is another and allied difficulty, which is
much more serious. I allude to the manner in which
species belonging to several of the main divisions of
the animal kingdom suddenly appear in the lowest
known fossiliferous rocks. . . .*

*If the theory be true, it is indisputable that before
the lowest Cambrian stratum was deposited long
periods elapsed, as long as, or probably far longer
than, the whole interval from the Cambrian age to the
present day; and that during these vast periods the
world swarmed with living creatures. . . .*

*To the question why we do not find rich
fossiliferous deposits belonging to these assumed*

earliest periods prior to the Cambrian system, I can give no satisfactory answer. . . . The difficulty of assigning any good reason for the absence of vast piles of strata rich in fossils beneath the Cambrian system is very great.[11]

Not only are all the phyla represented in the Cambrian period, there have been no new phyla found since that time.[12] And too, the more that is learned of Cambrian fossils, the more apparent it is that the living forms at that age were every bit as complex as are present forms of life.[13] If the evolutionary hypothesis were true, just the reverse should exist. Cambrian fossils should have been far more generalized and far more simple than any of the forms of life today. We also should today have additional phyla. Rock classified as "Pre-Cambrian" would be teeming with fossil evidence.

Even Ernst Haeckel was forced to admit that evolution could not explain the geological record. He said, "We cannot shut our eyes to the fact that various groups have from time of their first appearance, burst out into an exuberant growth of modification of form, size and members, with all possible, and one might almost say, impossible shapes, and they have done this within a comparatively short time, after which they have died out no less rapidly."[14]

And of the sudden emergence of the human intellect, anthropologist Loren C. Eiseley said that the human brain "measured in Geological terms, appeared to have been surprisingly sudden," and that "this huge mushroom of a brain, . . . has arisen magically between night and morning."[15]

With the sudden appearance of the human brain came the suddenness of civilization. Author of the book *New Discoveries in Babylonia about Genesis*, P. J. Wiseman reports:

No more surprising fact has been discovered by recent excavation, than the suddenness with which civilization appeared in the world. This discovery is the very opposite to that anticipated. It was expected that the more ancient period, the more primitive would excavators find it to be, until traces of civilization ceased altogether and aboriginal man

*appeared. Neither in Babylonia nor Egypt, the lands
of the oldest known habitations of man, has this been
the case.*[16]

After an examination of the geological evidence, the
author of *Age and Origin of Man*, Fredrich Pfaff, concludes:

*(1) The age of man is small, extending only a few
thousand years. (2) Man appeared suddenly: the most
ancient man known to us is not essentially different
from the now living man. (3) Transitions from the
ape to the man, or the man to the ape, are nowhere
found. The conclusion we are led to is that the
Scripture account of man, which is one and self-
consistent, is true . . . This account of man we accept
by faith, because it was revealed by God, is
supported by adequate evidence, solves the
otherwise insoluble problems, not only of science and
history, but of inward experience, and meets our
deepest need. . . . The more it is sifted and examined
the more well founded and irrefrangable does it
prove to be.*[17]

Admittedly, a belief in the scriptures which teach that the
Supreme Intelligence, Our Eternal Father in Heaven, God,
created this earth and placed life upon it, requires faith. But
faith is not something that is bad, *per se*. A faith can be
positive or negative, it can be either true faith or a false and
blind faith. Blind faith, the type of faith characterized by
those who would accept something as truth which is
contrary to the experience of men and opposed to revealed
truth, is not only un-testable but cannot be demonstrated by
what the scriptures call "good works." A test, then, of
faith—to determine whether or not it is true or false—would
be the examination of the ultimate consequences of the faith.

The blind, false faith of the evolutionary hypothesis can
claim no benefit to mankind. The useless perpetuation of the
many theories of spontaneous generation, natural selection,
survival of the fittest, and the supposed proofs offered by
morphology, paleontology, or embryology have hardly
contributed to the store of world knowledge. They have, in
fact, attempted to close the doors to learning by proclaiming
that all the answers have been found—that life has evolved

from non-life, that it will continue as it has in the past, and that there is no use in looking further at the facts. It is a LAW, they say, so don't investigate it.

Biologist H. Bentley Glass, retired president of the American Association for the Advancement of Science, and vice-president of a large state university, declared, "The great conceptions, the fundamental mechanisms, and the basic laws are now known. For all time to come, these have been discovered, here and now in our own life time." He remarked, "The endless horizons no longer exist."[18]

On the other hand, true faith, the faith that is harmoneous with both the correct interpretation of the scriptures and the cautious conclusions of the sciences, becomes in itself a motivation to search.

That the process of evolution has always been in operation cannot be demonstrated from science. But on the other hand, it has been demonstrated fully above that life forms appeared suddenly, and were as complex as modern forms. Their sudden appearance can neither be denied or explained by evolution.

Since evolution is at variance with the conclusions of true science, it follows that evolution is something outside science. It must be considered a faith, a dogma, a surrealistic philosophy that resists the correct thinking of both science and religion. It is a Golden Calf.

Chapter Fifteen
CLOCKS AND CHRONOMETERS

Hendrik Van Loon told the story of a little bird that flies high up to the north, to a land called Svithjod, once every thousand years to sharpen its beak on a rock. The rock is a hundred miles high and a hundred miles wide. When the rock has become worn away by the bird, tells Van Loon, then a single day of eternity will have gone by.

Because it cannot be measured or comprehended by itself, time is the most elusive element which man can discuss. Time, as such, cannot be calculated. That can be measured are the relative positions of events in time. This is important to any study which depends upon the proper reckoning of time, especially evolution.

Evolution, depends vitally upon the proper conception of magnitudes of time. Enough time must be made available in the theory to allow for all the events to occur by the alleged slow processes of evolution.

Since time itself cannot be measured, but only events occurring in time, what events should be used? What indications of past events have remained by which time could be measured? The answer to these questions will in part be determined by our ability to perceive those events correctly, using the technological developments of our age which allow us to observe and verify or refute conclusions.

One geologist explained that each grain of sand and each minute crystal in the rocks about us is a tiny clock, ticking off the years since it was formed. We need complex instruments to read them, but the fact remains that they are true clocks or chronometers. The story they tell, numbers the pages of earth history.[1]

The study of chronometry, the measuring of time, is the story of conflicting theories, though each theory is ostensibly attempting to measure some part of the real world. The lithosphere (the rocks) may be judged to be a certain age whereas the atmosphere another. The hydrosphere (the waters of the earth) may be found to be one age, but the age of the biosphere (the world of living organisms) might well be found to be another. Although great strides have been taken recently, and even though new formulae may be worked out in the future, chronometry remains one of the most disturbing tools of both science and evolution.

Any theory must expect to be modified as time goes on. The learning of today will antiquate the learning of yesterday. But as honest men search for the truths of the universe, we can be assured that the current popularity of any single theory will in time be reduced to its lowest common denominator—the observation and verification of reality by the scientific method. The theories that are incorrect or inadequate to explain the scientific world must and will be discarded. They will be replaced by more accurate ones. The new theories may also prove to be erroneous, but they will in time be replaced by better ones until, hopefully, the ultimate truth will be found and a precise theory be found to explain it.

There are several methods open to the scientist by which he may measure time. He may look to historical records to measure time. He may count tree rings, count varves (the annual rings of sedimentation left in some lakes), measure the moisture of volcanic obsidian, determine the amount of friction caused by the oceanic tides, or calculate the annual salt content of the oceans. He may measure the rates of sedimentation, or calculate the radioactivity of crystal formations and other radioactive minerals. And though the conclusions of several of these methods can be "adjusted" to correspond to each other, more often than not a different age will be calculated by each method, not to mention the different ages assigned to differing samples of the same substance. And so what happens in practice is that each scholar or group of scholars chooses the method of dating that "squares" with what he believes to be true.

There is no harm in a scientists being eclectic about the method of chronometry he chooses. All the methods have something to offer, but, of course, all do not agree with any one theory of the origin or creation of the earth and the life thereon. Nor do all agree with the theory of evolution.

As mentioned, one of the methods open to scientists is the testimony of the *historical records*. But historical records go back no more than about 6,000 years. "The earliest records we have of human history go back only about 5,000 years," said the *World Book Encyclopedia*.[2]

Dr. W. F. Libby, who received the nobel prize for his work with radio-carbon dating reported,

> Arnold and I had our first shock when our advisers informed us that history extended back only for 5,000 years. . . . You read statements to the effect that such and such a society or archeological site is 20,000 years old. We learned rather abruptly that these numbers, these ancient ages, are not known accurately; in fact, the earliest historical date that has been established with any degree of certainty is about the time of the 1st Dynasty in Egypt.[3]

Authors Mark A. Hall and Milton S. Lesser tell, "The invention of writing, about 6000 years ago, ushered in the historic period of man. The time prior to 6000 years ago is known as the prehistoric period."[4] And the author of *Man: His First Million Years*, Ashley Montagu, discovered that "The earliest written language, Sumerian cuneiform, goes back [only] to about 3500 B.C."[5]

6,000 years is just not long enough for the theory of evolution. The slow, gradual, blind change that evolution requires, and the finite possibility of positive change occurring spontaneously simply could not have occurred in a mere 6,000 years. The theory of evolution is weighed against the evidence of historical records, and the weight of history is discarded. The evolutionary hypothesis wins. Evolutionists prefer to use other methods of dating.

A second method open to the scientists is the *counting of tree rings*. As we all are taught, trees generally grow a new ring of living cells every year. It is a simple matter to count the annual rings to determine the age of the tree. Dendrochronologists, specialists who study the growth of tree rings, tell us that this dating method reveals that the oldest trees are not older than 5,000 years.

Dendrochronologist Edmund Schulman recently reported that "microscopic study of growth rings reveals that a bristlecone pine tree found [in 1957] at nearly 10,000 feet began growing more than 4,600 years ago. . . . Many of its neighbors are nearly as old; we have now dated 17 bristlecone pines 4000 years old or more."[6]

The assumption that trees consistently grow one ring every year is of course not true. Botanist Carl L. Wilson told, "The occurence of false growth rings may cause the age of the tree to be overestimated. Such rings are produced by a temporary slowing of growth during the growing season."[7] Botanist Wilfred W. Robbins reported that there are other factors, such as defoliation by insects, drought, and variation in the rainfall, that could also cause false growth rings or the absence of growth rings.[8] And botanists W. S. Glock and S. R. Agerter wrote,

> It has long been supposed that tree rings are formed annually and so can be used to date trees. The studies of tree ring formation . . . have shown that this is not always so, as more than one ring may be formed in one year.
>
> Two growth layers, one thick, the other thin and lenticular, proved to be more common than one growth layer in this particular increment. Three growth layers, in fact, were not unusual. A maximum of five growth layers was discovered in the trunks and branches of two trees.
>
> It must be pointed out that these intre-annuals were as distinctly and as sharply defined on the outer margin as any single annual increment.[9]

Dendrochronology has also been compared with the hypothesis of evolution. And as with the evidence of history, its evidence is also avoided. Evolution still stands unmarred.

Geologists and others have found that the deeper one goes into the earth the higher becomes the temperature of the earth. The deepest shaft into which man has descended is about five miles. On examining the earth's thermal history physicist Lord Kelvin theorized that the earth is cooling steadily and the age at which it once was molten was very recent. Kelvin had estimated the age of the earth to be at less than 100 million years. In 1897 he estimated the age of the earth at 24 million years.[10] His estimate that the earth is 20-40 millions is still too short a time for evolution to have occurred.

We should keep in mind that simply because there is time enough for an event to occur, it does not in any way imply that it did occur. Even if scientists were to demonstrate beyond any doubt (which a thing they are not capable of doing at the present) that the earth is 4 billion years old—a time in which, say the evolutionists, evolution could have occurred—that does not mean that evolution did occur.

Another chronometric method involves the measurement of the amount of salt in the oceans. Assumedly, the oceans were in the past millennia salt-free. The salt, or sodium chloride, which it now contains has been dissolved from the soil by the action of the rains forming streams then rivers which run into the oceans. Though the waters will evaporate, the salt remains. With the dissolving of more salt from the soils to run into the oceans, the salt content has gradually been built up to its present level. Calculating thusly, scientists have determined the age of the earth to be from about 180 million years to 338 million years old. If the oceans began with salt in them, the age of the earth would of course be over-estimated. And of course, the assumption that the oceans were originally salt-free cannot be proved.

Another method, and one of the earliest used to estimate geological time which did allow for evolution to have occurred, was the discovery of the stratified formations on the earth's crust, caused by sediment being carried into the oceans. Arthur Holmes in his book *The Age of the Earth* tells that the earth's stratified formations is at least 360,000 feet,[11] and that the annual discharge of sediments into the ocean would require millions of years. Just how many millions it must have taken has not been calculated with precision, but that the time involved is of the magnitude required by evolution makes this one of the more accepted dating methods.

In 1799 it was announced that sedimentary strata of the same age consistently showed the same types of fossils. After that announcement, the theory of the *"Geological Column"* has been the cornerstone of geology, the foundation of all systems of geologic dating. It was received warmly by the physical scientists, and more recently by the social scientists, not because it was and is without serious defect, but because it was sufficiently vague to allow for any amount of time required by the theory of evolution. One geologist fell into a pipe dream of idle splendor as he told that, "No longer was dogmatic creed to be superposed, by force if necessary, on facts denied by much material evidence. Centuries of supersitions fell before the challenging and rapidly advancing hypotheses about all manner of natural phenomena related to the planet earth and its relation to the universe."[12] This theory that fossils are laid down in a uniform pattern and in the same order, and are found always in the same kinds of soil appeared at first to warrant such enthusiastic fanfare, but upon investigation one finds glaring and insoluble problems.

Probably the most serious defect with the geological "column" is the basis on which it assigns geologic age to the stratum in the "column." Geologists begin by assuming that the theory of organic evolution propounded by Darwin and the theory of geologic evolution taught by Lyell are accurate. They then set about to examine the geological formations in which fossils were found. When they found a fossil which they esteemed as structurally simple, they said that the geological strata in which the fossil is found is old. If the fossil is relatively "specialized," that is to say, complex, then the strata in which it is found is to be considered new or recent. Thus, the age of the rock is determined by the complexity of the fossils it contains; and on the other hand, the age of the rock tells that the oldest fossils are the least complex. The circular reasoning of this argument is immediately apparent. Yale geologist Carl O. Dunbar pronounced, "Inasmuch as life has evolved gradually, changing from age to age, the rocks of each geologic age bear distinctive types of fossils unlike those of any other age. Conversely, each kind of fossil is an index or guide fossil to some definite geologic time. . . . Fossils thus make it possible to recognize rocks of the same age in different parts of the earth and in this way to correlate events and work out the history of the earth as a whole. They furnish us with a chronology, 'on which events are arranged like pearls on a string.' "[13]

Scientist Henry M. Morris commenting on this sort of reasoning, observed,

> There is a subtle example of circular reasoning here. Rocks are dated by the fossils they contain, rocks containing simple fossils being considered old and vice versa. This amounts simply to assuming as a prior fact that evolution is known to have occurred throughout geologic time. Then, the resulting geologic column, which its fossil series, is said to be the main, and indeed the only, proof that evolution has occurred.[14]

This is not the only difficulty with the "geological column." Though minor in comparison to the circular reasoning just mentioned, it is significant to note that geologists who use the "column" cannot determine the *absolute* age of either fossil or rock. At best all they can determine is the *relative* age of the specimen.

The relative geological ages are fairly well established, not by verification with the scientific method, but by consensus of the vocal majority, to be as follows:

Era	Epoc		Years Ago
CENOZOIC		Recent	0 - 10,000
		Pliestocene	1 million
	Teriary	Pliocene	15 million
		Miocene	30 million
		Oligocene	40 million
		Eocene	50 million
		Paleocene	60 million
MESOZOIC		Upper Cretaceous	80 million
		Lower Cretaceous	125 million
		Jurassic	160 million
		Triassic	200 million

PALEOZOIC	Permian	250 million
	Pennsylvanian	280 million
	Mississippian	310 million
	Devonian	350 million
	Silurian	410 million
	Ordovecian	470 million
	Cambrian	550 million
	Late Pre-Cambrian	1.6 billion
	Early Pre-Cambrian	2.7 billion
	Pre Crustal Rocks	4.5 billion

Even if we were to assume (which assumption would be well beyond what the scientific method could allow) that the geological column existed as outlined, the only means of placing the approximate dates when the certain epocs were supposed to have occurred is out and out guess-work. The methods are just not available to present-day science to determine these dates as closely as some insist them to be. Even if geologists could demonstrate that there were certian epocs beginning with pre-crustal rocks and continuing up to the recent one which began some 10,000 years ago, they would only be able to determine that what they call the Cambrian epoc was followed by the Ordovecian, which was followed by the Silurian, etc. You see, they are only able to say in what relative order the strata presents itself, and can say nothing about how long a time each stratum took to form itself.

Another problem with the column is that the column DOES NOT EXIST, except of course in the minds of some geologists. Now, when it is said that the column does not exist what is meant is that the entire column does not exist in any one place on the earth. It has been found here and there, a piece at this place, and another piece at that place. Sometimes the "column" is inverted, other times it is all mixed up. If the column were to exist in its totality, it would be over five miles thick! The Grand Canyon, the showcase of the geologists, is a mere mile deep, but the entire geological column would be five times deeper than that.

The geologic column presumes that life has evolved in an orderly progression and that as plants and animals in their varied evolutionary states died, they left their fossils in the stratum of their own geologic age. This assumption, of course, has invited a great deal of criticism.

Walter F. Lammerts reported that there are "over 500 cases that attest to a reverse order" of the geologic column, "that is, simple forms of life resting on top of more advanced types."[15]

The "geological column," like the theory of evolution itself, is a myth. It is a false religious faith shrouded in scientific jargon. It is an assumption the basis of which is another equally tenuous assumption. And the imprecision which the experts of the craft adjudge the age of the earth when they use the method, demonstrates its incredulity as a scientific tool. The following were the more prominent estimations of the age of the earth to the turn of the century based on estimates of the maximum thickness of sedimentary rocks:

Year	Geologist	Age in Millions of Years
1860	Phillips	96
1869	Huxley	100
1871	Haughton	1,526
1878	Haughton	200
1883	Winchell	3
1889	Croll	72
1890	de Lapparent	90
1892	Wallace	28
1892	Geikie	73 - 680
1892	McGee	1,584
1893	Upham	100
1893	Walcott	45 - 70
1893	Reade	95
1895	Sollas	17
1897	Sederholm	35 - 40
1899	Geikie	100
1900	Sollas	26.5
1908	Joly	80
1909	Sollas	80

When the experts disagree, whose opinion shall we choose? Shall we average out the opinions? Choose the latest estimate? Or is it possible that they are all working from a false premise, that they are all incorrect? The testimony of competent geologists and geophysicists proclaim their current difficulties. Geophysicist Arthur Beiser tells,

> While knowledge of the earth's size and shape is as ancient as Greek geometry and as modern as [Cape Kennedy's] rockets, man's understanding of the planet's origin—and its exact composition—is notoriously imprecise. . . . There are many more hypotheses than there are continents—nearly as many as there are geologists.[16]

J. H. F. Umbgrove said,

> But why should we not enter [the field of geology] if everyone who wants to join us in our geopoetic expedition into the unknown realm of the earth's early infancy is warned at the beginning that probably not a single step can be placed on solid ground?[17]

The periodical *Science Problems* explains why geochronometry is almost entirely wishful thinking:

> There is no single place on the earth where all the rocks in this series can be found. In any one place, some of them have been destroyed. Geologists studied the best examples of rocks in many places. Then, after long and patient work, they pieced the series of rocks together.[18]

At the introduction to the book *Outlines of Geology*, the author offered this wise caution:

> Much of the information contained in this book is within the well-lighted zone of proved fact. But no one ought to embark upon a study of even the elements of geology without realizing that we quickly pass from fact into a twilight zone or inference in which we can say, not "this is true," but only "probably this is true," and that thence we pass into

> a region of darkness lit here and there by a guess, a speculation. Speculation is a legitimate . . . thought process just as long as the thinker fully realizes that he is only speculating. But when he speculates, and at the same time persuades himself (and also, alas, his listeners) that he is drawing sound inferences, then knowledge does not progress. The reader of this book should remember at every page . . . what "we do not know" must be said or implied at nearly every turn, and finally that **what we do not know at present** would fill an indefinite number of volumes[19]

Further, evolutionist Le Gros Clark is forced to admit,

> There is one other source of misunderstanding in discussions on hominid evolution to which I should like to draw attention; it has reference to the importance of the time factor in the interpretation of fossil remains. In the past it sometimes happened that a great antiquity was assigned to the skeletal remains of **Homo sapiens** on the basis of what we now know to have been quite inadequate geological data.[20]

We have in this chapter discussed a few of the "clocks," or chronometers, which the scientists use to determine the age of the earth or any one part of it. We will discuss several other "clocks" in chapter sixteen. But it is imperative that the evidence of science be kept separate from the inferences which are made concerning the evidence, just as much as evidence must be distinct from testimony. We have found time and again that what an observer says must be the case in science does not mean that that is the way it is. If science is going to continue to broaden the horizons of human knowledge scientists must not be bogged down with unproved, and unprovable theories. They must never allow themselves the academic posture of claiming to know everything about the universe.

There is much to know, all the facts are not in yet. At best all that can be offered is a knowledgeable guess as to what has happened in the past, what time periods have elapsed, and the method or methods of creation of the life forms we know.

Chapter Sixteen

RADIOACTIVE CHRONOMETERS

Rontgen had discovered X-rays during the latter part of 1895 but it was still startling to M. Henri Becquerel (1852-1908) when in 1896 he found that a uranium salt that had been inadvertently left lying upon an unexposed photographic plate became exposed because of the mysterious radiation from his uranium sample. Further investigations by Becquerel, a French physicist, and his contemporaries soon led physics down a road that had never been previously traveled.

After Becquerel's discovery, Pierre and Marie Curie soon began their now famous but fatal experimentation which lead to the isolation of radium in 1903 and the discovery of other radioactive elements. And it was not long after the Curies that physicists discovered that the amount of radioactivity in a given sample would decrease at a specific rate. It "begins" presumably at 100%, then decreases to 50% at the end of what is called the element's "half-life," then decreases to 25% of its original radioactivity, during the same "half-life" period, then 12.5%, then 6.25%, then 3.125%, etc. All elements, as we shall find later, do not have the same half-life. Theoretically, any elements could be made radioactive, but with a few exceptions only those elements whose nuclei contain 82 or more protons are naturally radioactive. Physicists sometimes identify elements by the number of protons in the nucleus, which identity they call the "Z" number. Lead, for example, has a "Z" number of 82, and is naturally radioactive. Bismuth, 83; Polonium, 84; Astatine, 85; Radon, 86; Francium, 87; Radium, 88; Actinium, 89; Thorium, 90; Proactinium, 91; and Uranium, 92; are all found to be radioactive naturally. Some other elements (Thallium, for example, with a "Z" number of 81) are also radioactive naturally.

Radioactivity occurs when too many protons are packed together with neutrons in the nucleus and the binding forces are close to their limit of being able to hold the assemblage together. When the normal vibrations within the structure occasionally exceed the limit of a bond, then part of the nucleus spontaneously flies off.

One of the most usual particles of the nucleus to be thrown off is an *alpha particle*, which is composed of two neutrons and two protons. An alpha particle is the same as the nucleus of a helium atom. But the loss of the alpha particle leaves the nucleus in an exicted state and it does not "settle down" until it has released one or more *gamma rays*. A nucleus can also settle down, reach stability, or reach what is termed "ground state," by a transformation in which the nucleus changes its charge by one and emits an electron, together with a *neutrino*. The electron that is emitted is known as a *beta particle*.

Important to our discussion of radioactive chronometers is the fact that the nucleus may also reach stability by capturing anassociated electron and emitting a neutrino, which also changes the atom's charge by one. This is referred to as the *K-electron capture*. This method of transformation within the nucleus provides the basis for the potassium-argon method of dating.

Physicist Henry Faul explains the phenomenon of radioactivity this way:

> A radioactive nucleus may decay in a number of ways. It may emit an alpha particle, which consists of two protons and two neutrons bound tightly together. The alpha particle is identical with the nucleus of helium and is thrown out by some radioactive nuclei as a block with high velocity. In alpha decay, the atomic number, Z, of the nucleus decreases by two because of the removal of two protons, and the mass number, A, decreases by four. . . .

Another form of radioactive decay is beta emission. A beta particle is an electron, usually with negative charge, with is expelled by the nucleus in the process of decay. Nuclei do not contain electrons as such, but we may visualize the original of this electron as the decay of a neutron into a proton and an electron, with the emission of the electron as the beta particle. In this way, the number of protons in the beta decaying nucleus increases by one, and the parent nucleus becomes the nucleus of the next higher element in the periodic table. The parent and daughter in beta decay are, therefore, isobars. Isobars are nuclides of the same atomic mass, Z, but different atomic numbers; whereas isotopes are nuclides of the same atomic number, Z, but different masses.[1]

When a radioactive element, called the "parent" element decays it transforms itself into a new element, called the "daughter." For example, the parent Potassium-40 decays to the daughter element Argon-40, Uranium-235 becomes Lead-207, and the parent Uranium-238 becomes the daughter Lead-206. A chart showing the naturally occurring elements, the parents and their daughters, indicating their half-life or rate of decay and their type of decay shows some 21 elements that have been found to be radioactive:

Parent	Daughter	Half-life in Years	Type of of Decay
Potassium-40	Argon-40	1.2×10^9	Electron capture
Vanadium-50	Calcium-40	abt 6×10^{15}	Beta
	Titanium-50	abt 6×10^{15}	Electron capture
Rubidium-87	Chromium-50	4.7×10^{10}	Beta
	Strontium-87	4.7×10^{10}	Beta
Indium-115	Tin-115	5×10^{14}	Beta
Tellurium-123	Antimony-123	1.2×10^{13}	Electron capture
Lanthanum-138	Barium-138	1.1×10^{11} total	Electron capture
Cerium-142	Cerium-138	5×10^{15}	Beta
	Barium-138	5×10^{15}	Alpha
Neodymium-144	Cerium-140	2.4×10^{15}	Alpha
Samarium-147	Neodymium-143	1.06×10^{11}	Alpha
Samarium-148	Neodymium-144	1.2×10^{13}	Alpha
Samarium-149	Neodymium-145	abt 4×10^{14}	Alpha
Gadolinium-152	Samarium-148	1.1×10^{14}	Alpha

Dysporsium-156	Gadolinium-152	2×10^{14}	Alpha
Hafnium-174	Ytterbium-170	4.3×10^{15}	Alpha
Lutetium-176	Halfnium-176	2.2×10^{10}	Beta
Rhenium-187	Osmium-187	2×10^{10}	Beta
Platinum-190	Osmium-186	7×10^{11}	Alpha
Lead-204	Mercury-200	1.4×10^{17}	Alpha
Thorium-232	Lead-208	1.41×10^{10}	6 alpha, 4 beta
Uranium-235	Lead-207	7.13×10^{8}	7 alpha, 4 beta
Uranium-238	Lead-206	4.51×10^{9}	8 alpha, 6 beta

We should note in passing that some parent elements do not decay immediately to the daughter element. For example, Uranium-238 becomes Thorium-234, then becomes Proactinium-234 which decays to Uranium-234, which becomes Thorium-230, later to become Radium-226, then Polonium-218, then Lead-214, and then Lead-214 decays to Bismuth-214. Bismuth decays to either Thallium-210 then Lead-210 or Polonium-214 then Lead-210. Lead-210 becomes Bismuth-210, then Polonium-210, then Lead-206.

Of these radioactive elements only four are currently used to measure time: Uranium-235, Uranium-238, Rubidium-87, and Potassium-40. All the other decay too slowly or too rapidly, or are too rare in nature to be of much help in radiochronometry.[3] The following chart shows the radioactive parent-daughter sample and the minerals or rocks in which the radioactive materials are most commonly found:

Parent-Daughter	Minerals and Rocks
Uranium-238/Lead-206	Zircon Uranite Pitchblende
Uranium-235/Lead-207	Zircon Uranite Pitchblende
Potassium-40/Argon-40	Muscovite Biotite Hornblende Gauconite Sanidine Whole volcanic rock

Rubidium-87/Strontium-87	Muscovite
	Biotite
	Lepidolite
	Microcline
	Glauconite
	Whole Metamorphic rock

So far we have discussed something of radioactivity, half-life, the types of decay, and the sources of radioactive materials in nature. But how does radioactivity measure time? In what way can radioactivity be a clock? Nuclearphysicist Henry M. Faul explained that,

> When a radioactive nuclide is produced by some nuclear reaction and this production continues at a constant rate, [then] the amount of this new radioactive nuclide in a system gradually builds up to a constant value as the system approaches secular equilibrium. (The system in question may be a small crystal, or the whole Earth, or anything in between.) When a nuclide is in secular equilibrium, it is being produced in its environment just as fast as it decays, and therefore the amount remains constant.[4]

We might compare the radioactive equilibrium that exists in a sample to the filling of a bathtub with water when the drain plug is open. The water flows into the tub at a constant rate, the water drains out of the tub at a constant rate; but at a certain stage the amount flowing in equals the amount draining out. This is what is meant by equilibrium. The rate of formation of radioactivity in a sample is constant, and the half-life rate of decay of radioactivity is also constant. At a certain point the constant rate of formation is at equilibrium with the constant rate of decay.

When radioactivity is at equilibrium we may say that our "clock" is set at "zero." But once the rate of formation of the radioactive element ceases, then the decay rate slowly ticks away the time from the state of equilibrium.

Probably the most popularly discussed radioactive clock today is Carbon-14. Carbon-14 is produced from Nitrogen-14 high in the upper atmosphere by the action of cosmic rays. Chemically, the cosmic rays produce high-speed neutrons which collide with nitrogen atoms, knocking a positive proton from the nucleus of the nitrogen atom and replacing it with an uncharged neutron. Thus, Carbon-14 is formed.

The radioactive Carbon is diffused through the atmosphere, making up a minute quantity of the total carbon dioxide in the atmosphere. It is absorbed in plant tissue during the process of photosynthesis. The plants are then eaten by man and the animals, thereby diffusing Carbon-14 throughout all living things.

The rate of formation and ingestion of Carbon-14 is said to be at equilibrium with its decay rate, but at the death of the plant, animal, or man the rate of formation within that organism ceases. It is like turning the water "off" that flows into the tub. The open drain in the bottom of the tub, just as the natural half-life of the Carbon-14, allows a gradual reduction of the radioactive level. For Carbon-14 the half-life is 5,730 years by beta emission back to Nitrogen-14.[5]

The development of the radio-carbon dating method by W. F. Libby in the late 1940's was a virtual "atomic" bomb by which many archeologists saw their impending doom. Said Frederick Johnson in an article for the periodical *Science*, entitled "Radiocarbon Dating and Archaeology in North America,"

> With few exceptions. . . [archaeological dating] was by inference and guessing. . . . Libby's provision of a means of counting time—one that promised a definable degree of accuracy and worldwide consistence—caused all sorts of consternation because many of the new findings threw doubt on the validity of some established archaeological opinions.[6]

Johnson quotes one archeologist who reportedly said, "We stand before threat of the atom in the form of radiocarbon dating. This may be the last chance for old-fashioned, uncontrolled guessing."[7]

And though there is little doubt that radiocarbon dating, as with all the radiochronometric methods, is far superior to the guesswork of the traditional methods, these newer, more sophisticated methods are not without their shortcomings.

Nuclearphysicists Kunihiko Kigoshi and Hiroichi Hasegawa tell us that the basic assumption of atmospheric equilibrium in the formation/decay rate for Carbon-14 is now seriously questioned. They reported,

> An assumption on the constancy of atmospheric radiocarbon concentration in the past is basic for radiocarbon dating. However, the atmospheric radiocarbon concentration depends on the production rate of radiocarbon by cosmic rays in the stratosphere and the carbon cycle on the earth, and there is no evidence that either was constant in the past.[8]

It was recently reported in Science magazine that,

> Although it was hailed as the answer to the pre-historian's prayer when it was first announced, there has been increasing disillusion with the method because of the chronological uncertanties (in some cases, absurdities) that would follow a strict adherence to published C-14 dates. . . .
> What bids to become a classic example of "C-14 irresponsibility" is the 6000-year spread of 11 determinations for Jarmo, a pre-historic village in northeastern Iraq, which, on the basis of all archeological evidence, was not occupied for more than 500 consecutive years.[9]

It has also been told that "Errors of shell radiocarbon dates may be as large as several thousand years."[10] Science Year reported, "Scientists have found that the C-14 concentration in the air and in the sea has not remained constant over the years, as originally supposed."[11]

What of the dates then that have been arrived at by the Carbon-14 method? One report indicated that,

> It most certainly would ruin some of our carefully developed methods of dating things from the past. . . .
> If the level of carbon-14 was less in the past, due to a greater magnetic shielding from cosmic rays, then our estimates of the time that has elapsed since the life of the organisms will be too long.[12]

Libby himself was aware of his somewhat tenuous assumption. In 1955 he wrote,

*If one were to imagine that the cosmic radiation
had been turned off until a short while ago, the
enormous amount of radiocarbon necessary to the
equilibrium state would not have been manufactured
and the specific radioactivity of living matter would
be much less than the rate of production calculated
from the neutron intensity.*[13]

And more recently, geophysicist Richard E. Lingenfelter,
in 1963 wrote,

*There is strong indication, despite the large errors,
that the present natural production rate exceeds the
natural decay rate by as much as 25 per cent. . . . It
appears that equilibrium in the production and decay
of carbon-14 may not be maintained in detail.*[14]

And in 1965 Hans E. Suess, also a geophysicist, said,

*It seems probable that the present-day inventory of
Natural C*[14] *does not correspond to the equilibrium
value, but is increasing.*[15]

Though there are serious defects with Carbon-14
radiochronometry, at least it is a giant step in the right
direction towards putting the dating methods of archeology,
anthropology, and geology in the realm of the scientific
method. At last the scientsts are able to come to scientific
grips with the problems of dating, and can see clearly ahead
enough to define their direction and obstacles.

There are other radiochronometric methods, in addition to
Carbon-14, which deserve at least passing notice. But they
too have their drawbacks which are noted.

Spontaneous-fission Clock

By counting the fission tracks of Uranium-238 on a
crystal, the age of the crystal may be known. But the tracks
could have been produced by cosmic-ray interactions; as
well as spontaneous-fission of Uranium-235, or Thorium-
232. The evidence points to a very favorable chronometric
device, but further experimentation must be accomplished
first.

Lead-Lead Clock

The present-day ration of Uranium-235 and Uranium-238,
and their decay constants are known. Assumedly, therefore

only the ratio of their lead isotopes need be measured to determine an age. But unfortunately, there is indication that the system is not "closed," that is U-235/U-238 may allow the addition of more recent lead into the system.

Pleochroic Halos

Zircons and other small radioactive crystals are often found inside much larger crystals of mica, especially Biotite, and most of the alpha particles from the Uranium and Thorium in the Zircon spend most of their energy in the mica. As a result, a halo of radiation damage forms around the zircon crystal. The halos can be observed under polarized light, and measured with microphotometers. One difficulty with this dating method is that there is evidence that halos can be caused by forces external to the crystal.

Helium Clock

Once thought to be a very promising method of dating, the Helium clock has been found not to be a closed system.

Gross Uranium-Lead Clocks

Gross Uranium-lead clocks were among the first used to determine age by means of radioactivity. But it soon became evident that because of the high-mobility of the Uranium samples, and the fact that Uranium does not act well as a closed system, the method of dating was discontinued.

Ratio clocks from Ocean Sediments

Beryllium-10, Aluminium-26, Silicon-32, Chlorine-36, and Magnese-53—in addition to Carbon-14—are also produced by cosmic-ray bombardment in the atmosphere. There is also an estimated 10,000 tons of cosmic dust swept up by the earth in its orbit. But compared to Carbon-14 these elements are found in minute quantities. At present their extremely small amounts have prohibited accurate detection of their equilibrium activity. These elements could, at some future date be used in radiochronometry.

Tritium Clock

Hydrogen-3 (Tritium) is also produced by neutron bombardment in the atmosphere. It's half-life of only 12.3 years curtails its utility as a "clock." It cannot be used to measure the magnitudes of time necessitated by any theory of the past history of this earth.

Potassium/Argon

Carbon-14 radiochronometry can date nothing organic older than 50,000 years old, so what do the scientists use? According to an article in *Scientific American* "there is no way to date bone more than 50,000 years old, so they analyzed samples of rock from immediately above and below the level where the bones were found."[16] They analyze the Potassium-40/Argon-40 ration to determine the age of the rocks. They must assume that the fossil's age will be the same as that of the adjacent rocks.

The potassium of the earth naturally produces argon, but the argon can be boiled away" by volcanic action. This "boiling" action sets potassium/argon chronometer at "zero," thus making available to the scientists a radiochronometer. But there are some difficulties with this method. In the first place we start with an assumption that the volcanic activity removed even the most minute trace of argon—which it may not have done. Secondly, there is a possibility that the crystallizing rock containing the potassium could become contaminated from the atmosphere. In other words, the Potassium-40/Argon-40 chronometer may not be a closed system—a vital requirement in radiochronometry. And finally, the half-life of Potassium-40 is 1.3 billion years, and as such cannot date a substance with accuracy that is relatively young. As one scientific publication noted, "The [Potassium-Argon] dating method is increasingly inaccurate for dates of less than one million years. Consequently, there is a period during Early and Middle Pleistocene times when dating human remains is difficult and uncertain."[17]

Nine radiochronometric methods have been discussed briefly in this chapter. All nine methods require that certain assumptions be granted. And if all the assumptions are warranted, then the accuracy of the dating method would seem probable. But as has been shown, all of them have suffered from unwarranted assumptions. In conclusion, the assumptions are summarized:

Assumption 1: That the key to the past is the present. That the rates of formation and decay of radioactive elements was as it now is. That all environmental factors were essentially the same. That the present causes bring about the same effects as they did previously.

Assumption 2: That our measurement of the present is accurate.

Assumption 3: That our perception of events in the past is accurate. And specifically regarding radioactive chronometers.

Assumption 4: That the crystalline structure under experimentation is resistent to change during its entire lifetime.

Assumption 5: That a measurable quantity of some radioactive parent element is present.

Assumption 6: That a measurable amount of daughter element is present; and

Assumption 7: That there has been no loss or gain of either parent or daughter element throughout the lifetime of the crystal.

Those searching for truth by the scientific method, even as those searching for truth through religious means, must come to realize that all the facts are not in, that the processes are still dynamic, that man is still learning. The honest student of truth will find agreement with the words of scientist Merritt Stanley Congdon who said,

> Science is tested knowledge, but it is still subject to human vagaries, illusions and inaccuracies. . . . It begins and ends with probability, not certainty. . . .
>
> There is no finality in scientific inferences. The scientist says: "Up to the present, the facts are thus and so."[18]

Only as a dogmatist and not as a scientist could one say, "We have found the single key that unlocks all doors to all the mysteries of all creation."

Chapter Seventeen

THE FIRST
GREAT CAUSE

Over the years traditional religionists have expended reams of paper and hundreds of hours discussing what they called "The First Great Cause." What they were discussing was the process, the method, and the act by which, as they had supposed, the earth and the galaxy were instantly created. They naively believed that an Omnipotent Supreme Intelligence had instantaneously created the earth and the universe out of nothing. Though their belief was neither good science nor good religion they persisted in it.

Historically, the idea of the First Great Cause can be traced back as far as Plato, who, though excluding the creative act by a God, insisted that it happened as an event. He is credited with having said,

> They say that fire and water, and earth and air, all exist by nature and chance . . . and that as to the bodies which come next in order?earth, the sun, and the moon, and stars—existences. . . . After this fashion and in this manner the whole heaven has been created . . . not by the action of mind, as they say, or of any God, or from art, but as I was saying, by nature and chance only.[1]

And about the time of Jesus Christ the pagan Diodorus of Sicily wrote:

> Now as regards the first origin of mankind two opinions have arisen among the best authorities both in nature and history. One group, which takes the position that the universe did not come into being and will not decay, has declared that the race of men

*also has existed from eternity, there having never
been a time when men were first begotten; the other
group, however, which holds that the universe came
into being and will decay, has declared that, like it,
men had their first origin at a definite time.*[2]

Today the scientific world is also split. One group tells us
that our galaxy was created in an event which occured
about 4.5 billion years ago by a gigantic explosion; the other
says that though the elements of which this earth and these
galaxies are presently composed have always existed
(matter cannot be created nor destroyed, we are told, though
it may be transformed into energy) new stars and "earths"
are constantly being formed; yet a third says nothing about
origins, but affirms that the universe is in a continual state
of expansion and contraction. We shall discuss each of these
theories of the origin of the universe, pointing out briefly
their basic assumptions and perhaps their inadequacies.

One of the most remarkable early discoveries made with
the 100 inch telescope was that the galaxies appear to be
moving away from us as well as away from each other. This
observation gave rise to the "Expanding Universe" theory.
Since all the galaxies are receding from each other, it is
reasonable to suppose that at one time in the distant past
they were all in close proximity of each other. Assuming
that this was so, and based upon their calculated speeds
today they must have been crowded together about 3.5 to 4
billion years ago; thus the age of the universe can be
calculated.

A second method of dating the Universe is to observe star
clusters, such as the Pleiades, comprising some 200 stars. It
is assumed that these clusters must in time become
scattered by the gravitational forces operating in the
universe. It has been calculated that such clusters cannot
have remained together for more than 3 to 5 billion years,
and that if our galaxy were older than that, they would no
longer be found together. Operating under this premise, our
galaxy cannot be older than 5 billion years old.

A third basis for thinking the Universe is not older than
that is the fact that several of the stars we observe are in
reality double stars. The star Sirius, the Dog Star, is
actually two stars rotating about each other in close prox-

imity to each other. It is believed that over time these double stars will gradually separate from each other, leaving fewer and fewer double stars in the heavens. The high proportion of double stars leads astronomers to believe that the Universe is but a few billion years old.

It is assumed by astronomers that stars generate energy by converting hydrogen into helium. A current theory is that when the star's supply of hydrogen nears exhaustion it swells up and becomes what is known as a "Red Giant." From the brightness and size of these Red Giants it can be calculated, presumably, at what rate the hydrogen is being used, thus the age of the star. Red Giants, the oldest stars in the skies, have been calculated to be less than 4 billion years old.

Getting down now to the origin of the Universe there are two theories which should be mentioned.

First, is a theory first suggested by the Belgian physicist Lemaitre, in 1931. He believed that the matter of the universe originated by some gigantic atomic explosion which reduced all matter to its elemental and atomic form. Obviously, the greatest defect of this theory is that it accounts for nothing. It is an attempt to explain the origin of matter, but the premise upon which the theory is built requires that matter already be present.

In 1952 George Gamow sophisticated Lemaitre's theory by suggesting that the universe must have started as a tightly compressed mass of neutrons, perhaps as a gas. And as the gas expanded under extremely high temperatures and pressures the neutrons were split up into electrons and protons. Astronomer W. E. Filmer explains,

> *Although this chaos of colliding particles may appear at first sight to be hopelessly intractable, it does not, in fact, involve anything but comparatively simple processes which have been studied in the laboratory. Experiments with such high speed particles during the past twenty years provide the necessary knowledge of what the probable result of a collision between any two particles will be, provided their speeds are known. The only difficulty lies in the amount of calculation necessary to discover what the*

> *final mixture of gases will contain, when the temperature has fallen too low for collisions to be effective in building atoms.*[3]

Gamow's theory, too, does not account for the creation of matter, but only the formation of already existing, or pre-existent, matter.

While Gamow's theory accounts for Red Giants, that is, stars whose hydrogen is nearly depleted, it took Fred Hoyle to conjecture the seeming continuous creation in the universe. Hoyle taught that as long as a star has a supply of hydrogen which can be converted into helium—thus forming its supply of energy—its internal temperature will remain high enough to keep it blown up, much the same as the old hot-air balloons were kept inflated. Once the supply of hydrogen is depleted, the temperature suddenly falls, creating a partial vacuum within the star. Consequently, the outer matter of the star rushes inward toward the center of the star, creating an extremely dense star, a "White Dwarf." Extremely violent explosions are called "Supernovae." Thus, according to Hoyle's theory, the present arrangement of the galaxy resulted in first an explosion—the conversion of hydrogen into helium—then an implosion.

If we are led to accept the first theory, that the galaxy originated by a gigantic explosion then we must assume that all stars in our galaxy are of the same age. And since the light from these stars which we see today was actually transmitted from the star some considerable time ago, it would follow that those stars farthest from earth whose light is just now reaching us, would appear younger than the closer stars. No such thing is found with any demonstrable consistency.

If we believe the second theory, that of continuous creation, then we must believe that the oldest galaxies would be the largest—those whose stars are farthest apart, Again, no such thing is consistently found.

Astronomer Harlow Shapley, though irreverent of the Bible, told the limitations upon an astronomer:

> In the beginning was the Word, it has been piously recorded, and I might venture that modern astrophysics suggests that the word was hydrogen gas.

In the very beginning, we say, were hydrogen atoms; of course there must have been something antecedent, but we are not wise enough to know that.

Whence came these atoms of hydrogen . . . what preceded their appearance, if anything?

That is perhaps a question for metaphysics. The origin of origins is beyond astronomy.[4]

Another scientist asked,

How did hydrogen itself come into being? We cannot beg the question by supposing that it has always existed.

Hydrogen is steadily being converted into other elements by processes that seem irreversible. In spite of this hydrogen is still the most abundant element in the Universe.

We must, therefore, suppose that it has a finite age, for if it has existed for an infinite time, it should all have been used up by now.[5]

Though some have spoken of the "First Great Cause," and others, no doubt, will continue, Astronomer William Bonner asked another searching question,

What happened before the expansion started? Our model does not tell us. . . . Einstein's equations break down altogether. . . .

It is for this reason that some people refer to the start of the expansion as the creation of the universe. In some unknown way, it is argued, the matter of the universe was created at this moment. . . . We need not try to trace history back before this event, because the universe, and indeed, time itself, did not then exist.[6]

George Gamow himself admitted,

The Big Squeeze which took place in the early history of our universe was the result of a collapse which took place at a still earlier era, and the present expansion is simply an "elastic" rebound which started as soon as the maximum permissible squeezing density was reached. [But] Nothing can be said about the pre-squeeze density era of the universe.[7]

Gamow discussed the "Big Squeeze," but Bonner asks,

> The question we have to answer, though, is what
> can have made the contraction slow down, cease, and
> change to expansion. We ask why the collapsing
> cluster should slow down, stop, and then fly outward
> again.
>
> At present we have no answer: no physical
> mechanism which would reverse the contraction has
> yet been discovered.[8]

James Coleman revealed a failing of astronomers (and, we suppose, others). He observed that, "Regardless of the various areas a particular astronomer may be investigating, his findings always support the same theory that he avowedly champions. It is as if various scientists had been preordained to discover only evidence which supports their favorite theory! One wonders, then, if there isn't a great deal of evidence going undiscovered just because of this situation."[9]

To get away from the pitfalls of the two theories just discussed, the "Big Bang" theory which requires a gigantic explosion, and the "Pulsating" theory which implies that there was both a big bang" and then a contraction, followed by another "Big bang" and then a contraction, followed by another "big bang,"—the Swedish physicist Oskar Klein claimed the existence of what he called "anti-matter." Klein explained the impact of his theory upon the traditional theories of astronomy. He told,

> Obviously, if antimatter exists on a large scale, the
> current theories of the history of the universe—the
> "big bang" theory and the "steady state" theory fall
> by the wayside.
>
> If the original nucleus had contained antimatter as
> well as matter, it would have annihilated itself; the
> big bang would have been a too big bang.
>
> We do not venture to say how the cloud of
> ambiplasma originated. . . . We simply assume the
> existence of the cloud and go on to show that by
> gravitation it would begin to contract very slowly.[10]

Zophar, a friend to the biblical ancient, Job, asked, "Canst thou by searching find out God?"[11] Can the finite understand the Infinite? Just how much can unaided man comprehend in the universe around him? The scholars are very free to admit their limitations. They openly tell that they do not know of the origin of matter (not that anyone trained in religion could, either). They accept its existence and then try to make some sense out of what they see.

Said James A. Coleman,

> Modern cosmology and cosmogony, like other branches of science, are concerned with investigating the laws of the universe. They do not attempt to answer questions relating to an Original Cause—that is. where the laws of the universe came from or how they came into being. When giving a lecture on the origin of the universe, a scientist usually finds it difficult to handle questioners who persistently demand to know where the material originally came from which now makes up the universe.[12]

Writer Lincoln Barnett tells,

> Cosmologists for the most part maintain silence on the question of ultimate origins. leaving that issue to the philosophers and theology.[13]

The author of the text, Essentials of Earth History, wrote,

> Since the problem of the ultimate origin of the universe may be beyond the reach of human science, it is better for us to commence our discussion with the assumption that certain arrangements of matter and space are already in existence.[14]

Astrophysicist Jesse L. Greenstein has said,

> It is a terrible mystery how matter comes out of nothing. . . . We try to stay out of philosophy and theology. but sometimes we are forced to think in bigger terms, to go back to something outside science.[15]

On the origin of matter and the universe, Sir Bernard Lovell, director of the Nuffield Radio Astronomy Laboratories wrote,

> Any answer lies outside the scope of scientific observation and theory and . . . the answer to the cosmological problem may well contain other factors than observational astronomy and theoretical cosmology.[16]

And Fred Hoyle declares,

> There is an impulse to ask where originated material comes from. But such a question is entirely meaningless within the terms of reference of science.
>
> Why is there gravitation? Why do electric fields exist? Why is the universe?
>
> There queries are on the par with asking where newly originated matter comes from, and are just as meaningless and unprofitable.
>
> If we ask why the laws of physics . . . we enter into the territory of metaphysics—the scientist at all events will not attempt an answer. We must not go on to ask why.[17]

But though scientists are not to ask themselves "Why," the question persists. And as has been observed, certainly the answer to the question "How" has not been satisfactorily given. Though the answer to the question "How" may never be known here in mortality, there are perhaps two reasons for the scientists not having found the answers to the question "Why." One reason is that it is almost expected of a Scientist to reject the idea of a God. William Bonner said,

> It is the business of science to offer rational explanations for all the events in the real world, and any scientist who calls on God to explain something is falling down on his job.
>
> This is the one piece of dogmatism that a scientist can allow himself![18]

It seems almost inconceivable that a scientist, whose life's dedication is the pursuit of truth, regardless of its source, would summarily reject at least the possibility that God could somehow have anything to do with the world that the scientist observes.

Another difficulty which is admitted only occasionally by scientists today is the fact that they believe, nearly to the man, that the evolutionary hypothesis is correct. In 1959 Sir Julian Huxley told the 2,500 delegates assembled at the University of Chicago for the Darwinian Centennial that, "The evolution of life is no longer a theory. It is a fact. It is the basis of all our thinking."[19] And today the prophecy has almost been fulfilled.

Evolution has become the basis of nearly all scientific thinking. Scientist Kahn, in his work, *Design of the Universe*, admitted,

> We are today under the spell of the evolutionary thinking begun 150 years ago by Kant and Leplace in astronomy, by Thomas Buckle and Herder in history, by Buffon, Lamark and Darwin in biology. . . . We the children of these generations automatically think in terms of evolution, assume that everything had a beginning, and that this beginning was "chaos." . . . The question now arises as to whether astronomical problems can be solved by evolutionary trains of thought.[20]

Ir-religious faith in evolution has reduced the learning of the world to the level of the dark ages. Science seems to be shrouded in the dark robes of this false religious order. Its priests have led the learned into corners from which there seem to be no return. In science today, as in the dark ages of yesterday, denunciation of the theories of evolution means ostracism from the community of scholars and the brand of heretic.

The punishment meted out by this false faith is for the most part too great for many scientists to bear. But a day will soon come when there shall be a renaissance, a rebirth of learning where men will once again pursue truth, unshackled to the false traditions which today hold them bound.

Chapter Eighteen

CIVILIZATION: THE ANTITHESIS OF EVOLUTION

When Charles Darwin returned from a five year tour of the world on *H. M. S. Beagle* in 1836 it was an easy thing for him to label this culture or that "primitive." Bqt like other observations Darwin made, this idea too must now be laid to rest. Some of the leading anthropologists of today no longer consider many of them "primitive" at all. They regard them as "wreckage" of greater civilizations or as the product of "retrogressive evolution"—both of which are at odds with the entire thrust of Darwin's theory.

Some Anthropologists are now saying that,

> Many of the so-called "primitive" peoples of the world today, most of the participants agreed, may not be so primitive after all. They suggested that certain hunting tribes in Africa, Central India, South America, and the Western Pacific are not relics of the Stone Age, as had been previously thought, but instead are the "wreckage" of more highly developed societies forced through various circumstances to lead a much simpler, less-developed life.[1]

The *Encyclopedia Britannica* reported,

> In the early days of paleoanthropological discovery, **H. neanderthalensis** was commonly assumed to represent the ancestral type from which **H. sapiens** derived. . . . But the accumulation of further discoveries made it clear that these apparently primitive features are secondary—the

result of a retrogressive evolution from still earlier types which do not appear to be specifically distinguishable from *H. sapiens*. . . . Thus, the specialized Neanderthal type of *Homo* seems to have been preceded by a more generalized type. The brain of the specialized type was, surprisingly, rather large, for the mean cranial capacity actually exceeded that of modern human races.[2]

Modern evolutionist Sir Julian Huxley said that evolution was "a one-way process, irreversible in time producing apparent novelties and greater variety, and leading to higher degrees of organization, more differentiated, more complex, but at the same time more integrated."[3] But the evidence of history actually refutes these claims. Lange, in his *Commentary on Genesis* told,

Among human tribes left to themselves, the higher man never comes out of the lower. Apparent exceptions do ever, on close examination, confirm to the universality of the rule in regard to particular peoples, while the claim as made for the world's progress, can only be urged in opposition by ignoring the supernal aids of revelation that have ever shown themselves directly or on the human path.[4]

Hilbrecht, author of *Recent Researches in Bible Lands*, explained that "the flower of Babylonian art is found at the beginning of Babylonian history,"[5] not at the end as would be expected of under the theory of evolution. Of Egyptian "evolution" Professor A. H. Sayce reported,

The earliest culture and civilization to which the monuments bear witness was in fact already perfect. It was full-grown. The organization of the country was complete. The arts were known and practiced. Egyptian culture as far as we know at present has no beginnings.[6]

At another place he said,

The older the culture, the more perfect it is found to be. The fact is a very remarkable one, in view of modern theories of development and of the evolution of civilization out of barbarism. Whatever may be the reason, such theories are not borne out by the *discoveries of archaeology. Instead of the progress one should expect, we find retrogression and decay. Is it possible the Biblical view is right after all and that civilized man has been civilized from the outset?*[7]

Telling further of the greatness of the first Egyptians, Wm. Flinders Petrie explained that "the Great Pyramid bears on its stones the marks of the solid and tubular drill, edged with stone as hard as diamond, and cutting one-tenth of an inch at a revolution, and showing no sign of wear. They had also straight and circular saws. The same building reveals scientific and astronomical knowledge equal in some respects to modern science."[8] Egypt was definitely not the product of evolution.

Of the Greeks, Lecky, in his *History of European Morals*, wrote,

Within the narrow limits and scant populations of the Greek states, arose men, who in almost every conceivable form of genius, in philosophy, in epic, dramatic and lyric poetry, in written and spoken eloquence, in statesmanship, in sculpture, in painting, and probably in music, attained the highest levels of human perfection.[9]

Galton said of the intelligence of the ancient Greeks, "The millions of Europe, breeding as they have for two thousand years, have never produced the equal of Socrates and Phidias. The average ability of the Athenian race is, on the lowest possible estimate, nearly two grades higher than our own."[10]

Historians also tell that the Sumarians, the "grandparents" to the Babylonians, were anything but degenerate. As a matter of fact they trace no such thing as a gradual or evolutionary development in their culture. "The Sumarian culture springs into our view ready-made," tells Robert M. Engberg in *The Dawn of Civilization, and Life in the Ancient East.*[11]

Of the Aegean culture, Sir Arthur Evans explained that "the whole story of the excavations is indeed a marvelous one. They revealed an advanced civilization, cut off in its bloom, . . ."[12]

Professor of History and Religion, Hugh Nibley summarized the evidence against the theory of evolution in history.

To those whose view of the world comes from questionnaires and textbooks, it seems incredivle that the early dynastic civilization of Sumer, for example, should be so far ahead of later cultures that "compared with it everything that comes later seems almost decadent; the handicrafts must have reached an astounding perfection." (A. Goetze, Hethiter, Churriter and Assyrer, Oslop 1936, p. 11.) It is hard to believe that the great Babylonian civilization throughout the many centuries in which it flourished was merely coasting, sponging off the achievements of a much earlier civilization which by all rights should have been "primitive;" yet that is exactly the picture that Meissner gives us in his great study. (Bruno Meissner, Babylonian and Assyrien, Heidelberg; 1926, p. 154f.) It is against the rules that those artistic attainments for which Egypt is most noted—the matchless portraits, the wonderful stone vessels, the exquisite weaving—should reach their peak at the very dawn of Egyptian history, in the pre-dynastic period, yet such is the case. It is in the earliest dynasties and not in the later ones, that technical perfection and artistic taste of the Egyptians in jewelry, furniture, ceramics, etc., are most "advanced." "Here is a very odd thing," a British authority recently commented, "in literature the best in each kind comes first, comes suddenly and never comes again. This is a disturbing, uncomfortable, unacceptable idea to people who take their doctrine of evolution oversimply. But I think it must be admitted to be true. Of the very greatest things in each sort of literature, the masterpiece is unprecedented unique, never challenged or approached henceforth." (I. A. Richards, quoted by A. C. Bouquet, Comparative Religion, Penguin Books, 1951, p. 37.) More impressive is the report of the Egyptologist Siegfried Schott: "Time and again in the

development of Egyptian culture the monuments of a
new epoch present something heretofore unknown in
a state of completely developed perfection. . . ."
Please note that we are only able to pass judgment on
those things which happen to have survived from
those remote ages. We assume that those people were
crude in primitive in all **other** things, until some of
those other things turn up and show them to be far
ahead of us. We must admit, for example, that the
stone chipping of certain paleolithic hunters has
never been equalled since their day; it so happens
that stone implements are all that have survived
from those people—have we any right to deny them
perfection of other things? Is there any reason for
supposing that their wood or leather work was
inferior? . . . If it would not take us too far afield, I
could show you that the dogma of the evolutionary
advancement of the human race as a whole is nothing
but an impressive diploma which the nineteenth
century awarded—**summa cum laude**—to itself.
Modern man is a self-certified genius who, having
pinned the blue ribbon on his own lapel, proceeds to
hand out all the other awards according as the
various candidates are more or less like him.

"Yes," I can hear you say, "but there must have
been a long evolution behind all these early
achievements." This is for you to prove, not assume,
if you are a scientist. What is certain to date is (a)
that their evolutionary background has not been
discovered, and (b) that there is no record of **subsequent** improvement through all these thousands of
years. So let the biologists talk of evolution; for the
historian of ancient chronology can only regret "the
influence of a theory of evolutionism which has been
dragged so unfortunately into the study of ancient
history." (P. van Meer, The Ancient Chronology of
Western Asia and Egypt, Leiden; Brill, 1947, p. 13.)[13]

The evidence of history, to the chagrin of the
evolutionists, is in direct contradiction to the faith
propagated by Charles Darwin and his disciples. Instead of
simple beginnings, the historian finds the glory of the age.

Instead of the simple, peasant-like communities, great civilizations are found. Evolutionists look for simplicity early, but find just the opposite. History just doesn't square with the theory of evolution.

Once thought to be the irrefutable proof of the theory of evolution, the science of languages, now, like history, has become a barrier to the theory. Darwin said that the people of Terra del Fuego were the lowest in the scale, so far as discovered, and their language correspondingly crude. But further investigation shows that they have 32,430 words; over twice as many as Shakespeare used.[14] "The language of some of the tribes of the Congo is described by a missionary as more complex than Greek." The oldest languages are consistently found to be the most complex.

Max Mueller, in his *Lessons on the Science of Language,* told, "There is one barrier which no one has yet ventured to touch,—the barrier of language. Language is our Rubicon and no brute will dare to cross it. . . . No process of Natural Selection will ever distill significant words out of the notes of birds and animals."[16]

Linguist Otto Jesperson taught,

> We find that the ancient languages of our family Sanskrit, Zend, etc., abound in very long words; the further back we go, the greater the number of sesquipedalia. We have seen how the current theory, according to which every language started with monosyllable roots, fails at every point to account for actual facts and breaks down before the established truth of linguistic history.[17]

Linguist J. Vendryes explained that there is "nothing of the primitive" in the most ancient languages. He wrote,

> Some languages have been proved to be older than others, and certain of our modern tongues are known to us in forms more than two thousand years old. But the oldest known languages, the "parent languages," as they are sometimes called, have nothing of the primitve about them. Differ though they may from our modern tongues, they only furnish us with an indication of the changes which language has undergone, they do not tell us how language originated.[18]

An article in Science News Letter not too long ago confirmed the position that degeneration or simplification of languages as a fact of history was contrary to the expectations of the theory of evolution. It told, in part,

> There are no primitive languages, declares Dr. Mason, who is a specialist on American languages. The idea that "savages" speak in a series of grunts, and are unable to express many "civilized" concepts, is very wrong. . . .
>
> "In fact, many of the languages of non-literate peoples are far more complex than modern European ones, Dr. Mason said. . . .
>
> Evolution in language, Dr. Mason has found, is just the opposite of biological ekolution. Languages have evolved from the complex to the simple.[19]

Faced with the evidence, evolutionist Ashley Montagu declared, "Many 'primitive' languages . . . are often a great deal more complex and more efficient than the languages of the so-called higher civilizations."[20]

The evolutionists have told that in the process of civilizing the ancestors to modern man, that each civilization passed through certain ages—the stone age, the bronze age, and the iron age. But archaeologists have found no stone age in Africa.[21] They have found that in the ruins of Troy the bronze age was below [or earlier than] the stone age.[22]

They have discovered that the early Egyptians used bronze; whereas the later Egyptians used stone tools.[23] The "ages" as found among the Chaldeans were all mixed together.[24] And Europe had the metal age while America was apparently still in the Stone Age.[25] Today's civilizations have all the ages together. Again and again the evidence falls against the theory of evolution—all to demonstrate the inadequacy of the theory.

Darwin's theory made history at the Darwinian Centennial in 1959 when it was voted to the status of a scientific law. And it is continually making scientific history by stubbornly resisting every suggestion to take an unbiased assessment of itself and cull out what has been shown to violate the evidence of both science and history.

It is inconceivable that evolutionists persist in their belief that evolution is a fact. They tell that the fossil record is the strongest evidence of evolution. Yet an examination of the actual fossil evidence shows it to be almost totally lacking or inconclusive. The evolutionists then take refuge behind embryology. Embryologists, on the other hand remind them that they find not proof of evolution, but refutation. The weakness of the argument from morphology too, shows that evolution must look elsewhere. Darwin's modern counterparts declare that the vast amounts of time necessary for evolution to take place are proof that it did; but a survey of the chronometric methods available to science today demonstrates that a theoretically accurate method of dating is not known. And now, with the evidence of history, to what will they turn to support their theory? They will certainly find a new approach, a new harbor for their belief, another city of refuge, a new sanctuary for their false faith.

Evolution expects to be treated as a science but it behaves like a philosophy. It is couched in scientific language but resists every attempt to be scientific. It is taught as a scientific truth, but it cannot stand the light of scientific criticism. A creation out of rings and trinkets, the theory of evolution is a Golden Calf.

Chapter Nineteen

UNIFORMITARIAN OR CATACLYSMIC EVOLUTION?

The fundamental principle upon which both Charles Lyell and Charles Darwin built their respective theories of geologic and biologic evolution was the assumption that all physical and biological process which exist today have always continued at the same rate at which they proceed today. Physical processes such as radiation bombardment, evaporation, erosion, shifting of the earth, and biological process such as rates of mutations, life expectancy, and chemical reaction rates—all, they believe, have always proceeded at their present rates—that their present rates have been uniform throughout time. Both Lyell and Darwin, in other words believe in Uniformitarian Evolution. Though it was probably James Hutton who was the champion of uniformitarianism, credit must also go to Lyell and Darwin. Without these men uniformitarianism would have suffered an ignominious death.

Writing about Lyell's book, *Principles of Geology*, published in 1830, Geologist Don. L. Eicher exclaimed with considerable enthusiasm, "Geological problems now could be solved by reference to natural laws still active and available for study in the real world about us instead of by reference to former, shadowy, mythical, or supernatural events."[1] According to Eicher, if events have happened in the past just as they now occur, then there would be no need for a Supreme Being.

Contrary to the theory of uniformitarianism which supposes that the present is the key to the past, the theory of cataclysmic change accounts for terrible and abrupt changes in the earth.

Newsweek magazine recently observed,

> Catastrophism is a fighting word among geologists. It is a theory based on divine intervention, and its adherents held that the history of the earth and the life on it were moved by a series of disasters inspired by God—the last one Noah's Flood. It was the major line of thought for a few decades last century, but a vigorous counterattack by the naturalists against the supernaturalists eventually pushed it aside.
>
> But now many geologists believe the counterattack may have been all too vigorous. In their haste to reject the hand of God, they have passed over some solid evidence that could help improve their understanding of geology and evolution. . . .
>
> There is evidence, for example, that great expanses have been inundated within a matter of days. Such catastrophes were often followed by explosive development of different forms of life.[2]

The Saturday Evening Post reported,

> One of these periods of wholesale destruction of life occurred at the end of the last ice age. . . . It was a natural disaster which, according to one writer, destroyed some 40,000,000 animals in North America alone. . . . In a few thousand years life on earth assumed a radically new aspect. . . . It is apparent that millions of animals once flourished in areas now bitterly cold. . . .
>
> This discovery challenged the fundamental principle of the system established by the nineteenth-century geologist, Charles Lyell. He supposed that geological processes in the past always proceeded at their present rates: processes such as rainfall, snowfall, erosion and the deposition of sediment. . . . There was a very marked acceleration of the rate of these geological processes during the last part of the ice age. Some factor must, therefore have been operating that is not operating now.[3]

Just what that factor was, the uniformitarians are not saying. They cannot accept the evidence from geology and at the same time maintain the integrity of their scholastic stance. They choose generally to ignore the evidence.

But the evidence just won't be ignored. Time and again, wherever geologists look, they find evidence of natural violence. Says *Science Year* of 1965, "The discovery of coal and fossil ferns in the Transantarctic Mountains, . . . was evidence of a warm climate in the past."[4] A paleontologist from the American Museum of Natural History tells,

> *Geology students are taught that the "present is the key to the past," and they too often take it to mean that nothing ever happened that isn't happening now. But since the end of World War II, when a new generation moved in, we have gathered more data and we have begun to realize that there were many catastophic events in the past, some of which happened just once.*[5]

Evidence of one catastrophic event in Alaska indicates the violence about which scholars speak. Many times in the process of mining for gold it is necessary to cut through a considerable amount of "muck." Describing the process and the contents of the "muck" is F. Rainey, of the University of Alaska:

> *Wide cuts, often several miles in length and sometimes as much as 140 feet in depth, are now being sluiced out along stream valleys tributary to the Tanana in the Fairbanks District. In order to reach gold-bearing gravel beds an over-burden of frozen silt or "muck" is removed with hydraulic giants. This "muck" contains enormousnumbers of frozen bones of extinct animals such as the mammoth, mastodon, super-bison and horse.*[6]

F. C. Hibben, University of New Mexico, explains,

> *Although the formation of the deposits of muck is not clear, there is ample evidence that at least portions of this material were deposited under catastrophic conditions. Mammal remains are for the most part dismembered and disarticulated, even*

though some fragments yet retain, in their frozen
state, portions of ligaments, skin, hair, and flesh.
Twisted and torn trees are piled in splintered masses.
. . . At least four considerable layers of volcanic ash
may be traced in these deposits, although they are
extremely warped and distorted.[7]

In the New Siberian Islands the same picture of violent
catastrophism is painted. D. Gath Whitley reports,

The soil of these desolate islands is absolutely
packed full of the bones of elephants and
rhinoceroses in astonishing numbers. . . . These
islands were full of mammoth bones, and the quantity of tusks and teeth of elephants and rhinoceroses,
found in the newly discovered island of New Siberia,
was perfectly amazing, and surpassed anything
which had as yet been discovered.[8]

Whitley further tells of the state of preservation of these
creatures:

The contents of the stomachs have been carefully
examined; they showed the undigested food, leaves of
trees now found in Southern Siberia, but a long way
from the existing deposits of ivory. Microscopic
examination of the skin showed red blood corpuscles,
which was a proof not only of a sudden death, but
that the death was due to suffocation either by gases
or water, evidently the latter in this case. But the
puzzle remained to account for the sudden freezing
up of this large mass of flesh so as to preserve it for
future ages.[9]

In a provocative article in The Saturday Evening Post,
Ivan T. Sanderson described what he called the "Riddle of
the Frozen Giants." He reported in part,

About one-seventh of the entire land surface of our
earth, stretching in a great swath around the arctic
ocean, is permanently frozen. . . . It . . . includes . . .
masses of bones or even whole animals in various
stages of preservation or decomposition. So much of
the last is there on occasion that even strong men
find it almost impossible to stand the stench when it
is melting. . . . The list of animals that have been
thawed out of this mess would cover several pages . . .

woolly mammoths . . . woolly rhinoceroses, horses . . .
giant lion as well as many other animals now extinct
and some which are still in existence, like the musk
ox and the ground squirrel. . . . A scientific expedi-
tion was sent by the National Academy of Sciences
from St. Petersberg [when] the Beresovka mammoth
was discovered. . . . This company built a shack over
the corpse and lighted fires within to thaw it out. . . .
The lips, the lining of the mouth and the tongue were
preserved. . . . Upon the last, as well as between the
teeth, were portions of the animals last meal, which
for some incomprehensible reason it had not had time
to swallow. This meal proved to have been composed
of delicate . . . grasses and—most amazing of all—
fresh buttercup flowers. . . .

This discovery, in one full swoop, just about
demolished all the previous theories about the origin
of these frozen animals and set at naught almost
everything that was subsequently put forward. In
fact, it presented a royal flush of new riddles. First,
the mammoth was upright, but it had a broken hip.
Second, its exterior was whole and perfect, with none
of its two-foot-long shaggy fur rubbed or torn off.
Third, it was fresh; its parts, although they started to
rot when the heat of the fire got at them, were just as
they had been in life; the stomach contents had begun
to decompose. Finally, there were these buttercups in
its tongue. Perhaps none of these things sound very
startling at first, but if you will examine them one at
a time, employing simple logic and good, common
horse sense, you would immediately find that they
add up to an incredible picture. . . . It was not only
frozen but perfectly so, and here is where we come to
the first of the more vital points. . . .

Here is a really shocking—to our previous way of
thinking—picture. Vast herds of enormous, well-fed
beasts not specifically designed for extreme cold,
placidly feeding in sunny pastures, delicately
plucking flowering buttercups, at a temperature in
which we would probably not even have needed a
coat. Suddenly they were all killed, without any

visible sign of violence and before they could so
much as swallow a last mouthful of food, and then
were quick-frozen so rapidly that every cell of their
bodies is perfectly preserved, despite their great bulk
and their high temperature. What, we may well ask,
could possibly do this. Fossils of plants requiring
sunlight every day of the year—which is far from the
condition pertaining about the poles—have been
found in Greenland and on Antarctica.[10]

The evidence of cataclysm is not confied to the arctic
regions. Immanuel Velikovsky, in his fully documented
work, *Earth in Upheaval*, tells of rocks, sometimes gigantic
in size, that had often been transported great distances. He
reports,

There are erratic boulders in many places of the
world. In the British Isles, on the shore and in the
highlands, are enormous quantities of them,
transported there across the North Sea from the
mountains of Norway. Some force wrested from
those massifs, bore them over the entire expanse that
separates Scandinavia from the British Isles, and set
them down on the coast and on the hills. From
Scandinavia boulders were also carried to Germany
and spread over that country, in some places so
thickly that it seems as though they had been
brought there by masons to build cities. Also, high in
the Harz Mountains, in central Germany, lie stones
that originated in Norway.

From Finland blocks of stone were swept to the
Baltic regions and over Poland and lifted onto the
Carpathians. Another train of boulders was fanned
out from Finland, over the Valdai Hills, over the site
of Moscow, and as far as the Don.

In North America erratic blocks, broken from the
granite of Canada and Labrador, were spread over
Maine, New Hampshire, Vermont, Massachusets,
Connecticut, New York, New Jersey, Michigan,
Wisconsin, and Ohio; they perch on top of ridges and
lie on slopes and deep in the valleys. They lie on the
coastal plain and on the White Mountains the

> Berkshires, sometimes in an unbroken chain; in the
> Pocono Mountains they balance precariously on the
> edge of crests. The attentive traveler through the
> woods wonders at the size of these rocks, brought
> there and abandoned sometime in the past,
> frighteningly piled up. . . .
>
> In innumerable places on the surface of the earth,
> as well as on isolated islands in the Atlantic and
> Pacific and in Antarctica, lie rocks of foreign origin,
> brought from afar by some great force. Broken off
> from their parent mountain ridges and coastal cliffs,
> they were carried down dale and up hill and over
> land and sea.[11]

During times of great catastrophe the seas were by no
means calm, as was reported by Georges Cuvier.

> It has frequently happened that lands which have
> been laid dry, have been again covered by the waters,
> in consequence either of their being engulfed in the
> abyss, or of the sea having merely risen over them. . . .
> These repeated irruptions and retreats of the sea
> have neither all been slow nor gradual; on the
> contrary, most of the catastrophes which have
> occasioned them have been sudden; and this is
> especially easy to be proven, with regard to the last
> of these catastrophes, that which, by a twofold
> motion, has inundated, and afterwards laid dry, our
> present continents, or at least a part of the land
> which forms them at the present day.[12]

The "last of these catastrophes" has been dated at about
five or six thousand years ago. Velikovsky explains that
when the catastrophe "covered with mud and pebbles the
bones in the Kirkdale cave" it was found that "the bones
were not yet fossilized; their organic matter was not yet
replaced by minerals." He then tells that "Buckland thought
that the time elapsed since a diluvian catastrophe could not
have exceeded five or six thousand years, the figure adopted
also by DeLuc, Dolomieum and Cuvier, each of whom
presented his own reasons."[13]

Professor of Geology at Oxford (1874-1888) Joseph
Prestwich believed that "the south of England had been
submerged to the depth of no less than about 1000 feet

between the Glacial—or Post-glacial—and the recent or Neolithic periods."[14]

Prestwich also told of the masses of fragmented bones of panther, lynx, caffir-cat, hyaena, wolf, bear, rhinoceros, horse, wild boar, red deer, fallow deer, ibex, ox, hare and rabbit have been found in the faults and fissures of the rock of Gibraltar. "The bones are most likely broken into thousands of fragments—none are worn or rolled, nor any of them gnawed, though so many carnivores then lived on the rock. A great and common danger, such as a great flood, alone could have driven together the animals of the plains and of the crags and caves," surmises Prestwich.[15]

In the Cumberland cavern, reported Velikovsky, were the same signs of a great deluge which forces all animal life to central locations, thereupon smashing them to bits. He tells,

> So also it happened that animals of northern regions—wolverine and leming, the long-tailed shrew, mink, red squirrel, muskrat, porcupine, hare, and elk—were heaped together with animals "suggesting warmer climatic condition"—peccary, crocodilid, and tapir. Animals that now live on the western coast of America—coyote badger, and puma-like cats—are in this assemblage. Animals that live in areas of plentiful water supply—beaver and muskrat and mink—are found in the Cumberland cavern jumbled together with animals of arid regions—coyote and badger—and those of wooded regions together with animals of open terrain, like the horse and the hare. This is truly "a peculiar assemblage of animals." Extinct animals are found intermingled with extant forms. Death came to all of them at the same time. Any theory that attempts to explain the presence of animal bones from various climates in one and the same locality by a sequence of glacial and interglacial periods must stumble on the bones of the Cumberland cavern.[16]

At one time or another in the very recent past some great catastrophe heaped sand into one area where before was herded cattle. The place: The Sahara Desert. "What is now the desert of Sahara was an open grassland or steppe in earlier days," tells Velikovsky. "Drawings on rock of herds of cattle, made by early dwellers in this region, were discovered by Barth in 1850. Since then many more drawings have been found. The animals depicted no longer inhabit these regions, and many are generally extinct."[17] Neolithic instruments have been found close to the drawings. It almost seems unbelieveable to learn that men pastured cattle on what is today the largest desert on earth—a desert which covers some 3.5 million square miles, an area as large as the entire continent of Europe!

Another cataclysm can be easily detected by what the geologists call "paleomagnitism." The geologists explain that molten rock is non-magnetic or loses its magnetic state when liquified; but will acquire the magnetic state and orientation of the magnetic field of the earth when it has cooled to 580 degrees centigrade.[18] This "paleomagnetism" acquired by the rock upon cooling will remain with the rock regardless of its subsequent relocation or reorientation toward the earth's lines of magnetism. That is to say, if an earthquake, for example, jarred a lava bed, breaking it up, sending rock in every direction, the lines of magnetism that the rock acquired upon cooling to 580 degrees would remain in the rock. Each rock would therefore become, in a sense, a magnet with a "north" and a "south" pole, and would possess a minute but often measurable amount of magnetism.

A definite problem for uniformitarian evolutionists is that it is fairly obvious that the earth's magnetic field has reversed itself. Geologist H. Manley explains his work with paleomagnetism this way:

> Sufficient experiments have now been made to allow only one plausible explanation of this "inverted" magnetization—that the Earth's magnetic field was itself reversed at the period when the rocks were formed.[19]

S. K. Runcorn, of Cambridge, confirms Manley's work, saying that, "The evidence accumulates that the earth did reverse itself many times."[20] "The north and south geomagnetic poles reversed places several times," he said, observing that "The field would suddenly break up and reform with opposite polarity."[21]